Lecture Notes in Computer Science　　10121

Commenced Publication in 1973
Founding and Former Series Editors:
Gerhard Goos, Juris Hartmanis, and Jan van Leeuwen

More information about this series at http://www.springer.com/series/7411

Tatiana K. Madsen · Jimmy J. Nielsen
Nuno K. Pratas (Eds.)

Multiple Access Communications

9th International Workshop, MACOM 2016
Aalborg, Denmark, November 21–22, 2016
Proceedings

 Springer

Editors
Tatiana K. Madsen
Aalborg University
Aalborg
Denmark

Nuno K. Pratas
Aalborg University
Aalborg
Denmark

Jimmy J. Nielsen
Aalborg University
Aalborg
Denmark

ISSN 0302-9743   ISSN 1611-3349   (electronic)
Lecture Notes in Computer Science
ISBN 978-3-319-51375-1   ISBN 978-3-319-51376-8   (eBook)
DOI 10.1007/978-3-319-51376-8

Library of Congress Control Number: 2016960730

LNCS Sublibrary: SL5 – Computer Communication Networks and Telecommunications

Printed on acid-free paper

This Springer imprint is published by Springer Nature
The registered company is Springer International Publishing AG
The registered company address is: Gewerbestrasse 11, 6330 Cham, Switzerland

Preface

It is our great pleasure to present the proceedings of the 9th International Workshop on Multiple Access Communications (MACOM) that was held in Aalborg during November 21–22, 2016. Previous editions were organized in Helsinki (2015), Halmstad (2014), Vilnius (2013), Maynooth (2012), Trento (2011), Barcelona (2010), Dresden (2009), and Saint-Petersburg (2008).

Our gratitude goes to the Technical Program Committee and external reviewers for their efforts in selecting ten high-quality contributions (out of 12 submitted) to be presented and discussed in the workshop.

The contributions gathered in this proceedings volume describe the latest advancements in the field of multiple access communications.

Finally, we would like to take this opportunity to express our gratitude to all the participants, together with the local organizers, who helped to make MACOM 2016 a very successful event.

November 2016

Tatiana Madsen
Alexey Vinel
Boris Bellalta
Jimmy Nielsen
Nuno Pratas

Organization

Executive Committee

General Chair

Tatiana Madsen   Aalborg University, Denmark

TPC Chairs

Alexey Vinel   Halmstad University, Sweden
Boris Bellalta   Universitat Pompeu Fabra, Spain

Local Chairs

Jimmy Nielsen   Aalborg University, Denmark
Nuno Pratas   Aalborg University, Denmark

Technical Program Committee

Konstantin Avrachenkov	Inria Sophia Antipolis, France
Abdelmalik Bachir	Biskra University, Algeria
Sandjai Bhulai	VU University Amsterdam, The Netherlands
Raffaele Bruno	IIT-CNR, France
Peter Buchholz	TU Dortmund, Germany
Claudia Campolo	University Mediterranea of Reggio Calabria, Italy
Cristina Cano	Inria Lille-Nord Europe, France
Tugrul Dayar	Bilkent University, Turkey
Alexander Dudin	Belarusian State University, Belarus
Marc Emmelmann	Fraunhofer FOKUS, Germany
Dieter Fiems	Ghent University, Germany
Andres Garcia-Saavedra	NEC Labs Europe, Germany
Andras Horvath	University of Turin, Italy
Ganguk Hwang	KAIST, South Korea
Dragi Kimovski	University of Innsbruck, Austria
Valentina Klimenok	Belarusian State University, Belarus
Kristina Kunert	Halmstad University, Sweden
David Malone	Maynooth University, Ireland
Arturas Medeisis	International Telecommunication Union, Saudi Arabia
Dmitry Osipov	IITP RAS, Russia
Evgeny Osipov	LTU Luleå University of Technology, Sweden
Edison Pignaton de Freitas	Federal University of Santa Maria, Brazil

Vicent Pla	Universitat Politecnica de Valencia, Spain
Taneli Riihonen	Aalto University School of Electrical Engineering, Finland
Zsolt Saffer	Budapest University of Technology and Economics, Hungary
Nikos Sagias	University of the Peloponnese, Greece
Pablo Salvador	IMDEA Networks Institute, Spain
Bruno Sericola	Inria Rennes - Bretagne Atlantique, France
Susanna Spinsante	Università Politecnica delle Marche, Italy
Andrey Trofimov	Saint-Petersburg State University of Aerospace Instrumentation, Russia
Bernhard Walke	RWTH Aachen University, Germany
Till Wollenberg	University of Rostock, Germany
Yan Zhang	Simula Research Laboratory and University of Oslo, Norway

Editorial

Tatiana Madsen, Jimmy Nielsen, and Nuno Pratas

Aalborg University, Aalborg, Denmark

The International Workshop on Multiple Access Communications (MACOM) aims to gather international researchers to discuss the most recent developments on multi-user communications theory and multiple access techniques, focusing on PHY and MAC layer protocols. In the ninth iteration of the MACOM, we assembled a strong program where the different papers tackle challenges within wireless cellular networks, wireless local area networks, power line communications, as well as more general contributions that are applicable to a large range of network architectures and systems.

In the paper "A Secrecy and Security Dilemma in OFDM Communications," A. Garnaev et al. discuss the trade-off between secrecy and security in an OFMD communications context. The authors propose the use of a Kalai–Smorodinsky bargaining solution between secrecy and security for OFDM-style communications, which allows one to adjust the selection of the security level of communication.

In "An Analytical Model for Perpetual Network Codes in Packet Erasure Channels;" P. Pahlevani et al. provide a new analytical model for the evaluation of perpetual network codes. Specifically, the authors show that perpetual codes introduce linear dependent packet transmissions in the presence of an erasure channel.

In "Generalized Minimum Distance Decoder in a DHA FH OFDMA Employing Concatenated Coding," D. Osipov et al. propose a novel decoder that is applicable to DHA FH OFDMA. By employing robust reception and concatenated coding, construction is considered. The problem of employing a generalized minimum distance (GMD) decoder as an outer code decoder is considered.

In "Modemless Multiple Access Communications over Powerlines for DC Microgrid Control," M. Angjelichinoski et al. describe a novel technique that allows for multiple access communications over powerlines without requiring modems, using the microgrid's controllers. The authors present the proposed solution in the context of the distributed optimal economic dispatch, where the generators periodically transmit information about their local generation capacity, and, simultaneously, using the properties of the multiple access channel, detect the aggregate generation capacity of the remote peers, with the aim of distributed computation of the optimal dispatch policy.

In "An Experimental Study of Advanced Receivers in a Practical, Dense, Small Cells Network," D. Wassie et al. present a detailed experimental study on the performance of advanced receivers in a dense network setting. The authors show, in a practical uncoordinated dense small cell deployment, that advanced receivers can alleviate the need for detailed cell planning.

In "On the Impact of Precoding Errors on Ultra-Reliable Communications," G. Pocovi et al. study the impact of precoding errors on ultra-reliable communications. The authors show that closed-loop microscopic diversity schemes are generally preferred over open-loop techniques to achieve the SINR outage performance required for ultra-reliable communications. Macroscopic diversity, where multiple cells jointly serve the UE, provides additional robustness against precoding errors.

In "Joint Usage of Dynamic Sensitivity Control and Time Division Multiple Access in Dense 802.11ax Networks," E. Khorov et al. evaluate the use of control techniques to adapt TDMA applied to the 802.11ax networks. The authors show that in dense deployments the hidden node problem is present and occurs even when the distances between access points and associated stations are small.

In "Generic Energy Evaluation Methodology for Machine Type Communication," T. Jacobsen et al. developed an energy consumption model applicable to machine-type communications protocols. The authors pay special attention to factors such as power control, link-level performance and a radio power model with a non-constant power amplifier (PA) efficiency model intended for massive machine type communication devices. The results show the impact of these factors and highlight that applying a commonly used constant radio PA efficiency model can result in an overestimation of the battery life of up to 100% depending on the traffic scenario.

In "Performance Evaluation of LAA-LTE and WiFi Coexistence in Unlicensed 5-GHz Band Under Asymmetric Network Deployments Using NS3," M. Nurchis et al. provide a detailed performance evaluation of the coexistence of unlicensed LTE with WiFi. Specifically, for LTE to fairly coexist with WiFi, many open issues still require investigation, as there is not yet a clear consensus on the actual impact that each network can have on the performance of the other. The authors show that in scenarios in which the two networks have unbalanced and asymmetric location, the access point and base station locations and their mutual distance play a significant role in the overall and individual performance.

Finally, in "Providing Fast Discovery in D2D Communication with Full Duplex Technology," M. Sarret et al. show that autonomous proximity discovery in a D2D context can be greatly enhanced using full duplex transceivers, which can transmit and receive simultaneously in the same band. The presented topics provided an overview of the current open research topics in multiple user communications.

Contents

Information Theory

Physical Layer Aspects

An Experimental Study of Advanced Receivers in a Practical Dense Small Cells Network

Dereje A. Wassie[1(✉)], Gilberto Berardinelli[1], Fernando M.L. Tavares[1],
Troels B. Sørensen[1], and Preben Mogensen[1,2]

[1] Department of Electronic Systems, Aalborg University, Aalborg, Denmark
{daw,gb,ft,tbs,pm}@es.aau.dk
[2] Nokia Bell Labs, Aalborg, Denmark

Abstract. 5G is targeting a peak data rate in the order of 10 Gb/s and
at least 100 Mb/s data rate is generally expected to be available everywhere. For fulfilling such 5G broadband targets, massive deployment of
small cells is considered as one of the promising solutions. However, inter-
cell interference leads to significant limitations on the network throughput in such deployments. In addition, network densification introduces
difficulty in network deployment. This paper presents a study on the benefits of advanced receiver in a practical uncoordinated dense small cells
deployment. Our aim is to show that advanced receivers can alleviate
the need for detailed cell planning. To this end we adopt a hybrid simulation evaluation approach where propagation data are obtained from
experimental analysis, and by which we analyse how MIMO constellation and network size impacts to the aim. The experimental data have
been obtained using a software defined radio (SDR) testbed network with
12 testbed nodes, configured as either access point or user equipment.
Each node features a 4×4 or a 2×2 MIMO configuration. The results
demonstrate that advanced receivers with a larger MIMO antenna configuration significantly improves the throughput performance in a practical
dense small cells network due to the interference suppression capability.
In addition, the results prove that the operators can rely on uncoordinated deployment of small cells, since the resulting interference can be
suppressed by the advanced receiver processing with sufficiently capable
MIMO antenna configuration.

1 Introduction

Currently, users located in indoor environment consume around 80% of all the
mobile broadband traffic [1]. In addition, the data traffic demand is expected
to increase at a tremendous pace year by year due to the proliferation of new
services, broadband applications, rapid adoption of smartphones & devices and
so on. To attain the demands, industry and academia are spending a significant
effort on the design of a new 5th Generation (5G) radio access technology (RAT).
5G is expected to accommodate a wide range of diverse services, including new
mission critical communication services targeting ultra-high reliability and very
low latency transmission for e.g. industrial automation and vehicular to vehicular

© Springer International Publishing AG 2016
T.K. Madsen et al. (Eds.): MACOM 2016, LNCS 10121, pp. 3–14, 2016.
DOI: 10.1007/978-3-319-51376-8_1

communication, massive machine type of communication and enhanced Mobile Broadband (eMBB) services.

5G is targeting a peak data rate in the order of 10 Gb/s and at least 100 Mb/s data rate is generally expected to be available everywhere [2] with respect to broadband applications. For fulfilling such 5G targets, network densification is crucial and a massive deployment of small cells is foreseen [3]. However, network densification complicates the network planning and raises challenging issues for site selection and acquisition in indoor small cells scenario. Furthermore, in dense small cells networks, the throughput performance is mainly limited by inter-cell interference. With respect to that, an accurate site planning, i.e. selecting the locations and the number of cells, may reduce the impact of inter-cell interference and boosts the network throughput performance. However, this is very complicated in indoor environment due to a large number of walls and the different floor plans in each building. Therefore, operators are forced to make complex plans to find out where each cell should be installed in such a way that the interference is not a burden or operator can employ traditional inter-cell interference mitigation where the cells are configured to operate in different portions of the available spectrum. However, this approach will limit efficient usage of spectrum. In addition, the spectrum allocation will face a huge challenge due to a massive number of small cells. Alternatively, operators can employ uncoordinated deployment, which requires less planning and relies on advanced baseband processing at the receiver side to mitigate inter-cell interference.

The authors in [4] presented the envisioned 5G small cells concept in which the interference mitigation technique lies on the usage of the Interference Rejection Combining (IRC) receiver. The key working principle of the receiver relies on exploiting degrees of freedom of MIMO transceiver antenna for projecting the significant interfering signal over an orthogonal subspace of the desired signal to diminish the detrimental impact of inter-cell interference [5]. In addition, the envisioned 5G system frame structure design employs Time Division Duplex (TDD) mode, aiming to support efficient estimation of the interference covariance matrix, which is the key factor for the good performance of the IRC receiver [13]. Moreover, we have previously verified the potential of the IRC receiver towards boosting the network throughput in dense small cell scenarios based on both system level simulations [6] and experimental testbed networks [7].

In this paper, we address the potential benefit of the IRC receiver with different MIMO antenna configurations in a practical indoor dense small cells deployment and show its' inherent capability to cope with challenging interference conditions, and attain the best effort throughput improvement without the need for cell planning. We are adopting a hybrid simulation evaluation approach where the propagation models are replaced with actual measurement data obtained from a large testbed network. These field channel measurements provide the complete and factual information about all existing link combinations in a given deployment. The large testbed network consists of 12 software defined radio (SDR) testbed nodes which can be classified as 6 cells with one access point (AP) and one user equipment (UE), and the SDR nodes are equipped with a 4×4

MIMO antenna transceiver. Our previous work [7] on the experimental evaluation of advanced receivers considered a lower order MIMO antenna configuration and a limited number of cells. Here, we are assuming a larger, and more realistic, configuration of cells which can provide a further insight on the potential capability of the IRC receiver where the interference conditions become challenging in practical dense small cell deployments dominated by inter cell interference. We also examine the impact of MIMO constellation size in combination with different number of cells, which configured to operate on the same carrier frequency, with the aim of showing that advanced receivers can alleviate the need for planning.

The paper is organized as follows. Section 2 presents our multi-link MIMO channel sounder testbed setup, while Sect. 3 presents the channel measurement campaign in an indoor office environment. Section 4 discusses the experimental performance results of advanced receiver toward dealing with inter-cell interference in a practical uncoordinated dense small cells network. Finally, conclusions and future work are recalled in Sect. 5.

2   Multi-link MIMO Channel Sounder Testbed Setup

In this section, we present the multi-link MIMO channel sounder testbed. The testbed adopts Universal Software Radio Peripheral RIO (USRP-2953R) boards by National Instruments [9]. Each testbed node is composed of two USRP-2953R boards connected with PCI express cable to a host computer, as shown in Fig. 1. The USRP boards are used only as a radio-frequency front end, while the entire baseband processing runs on the host computer. The host computer runs a multi-link MIMO channel sounder application which is developed by using LabVIEW communications suite design environment [8]. Each USRP-2953R board has two radio frequency chains, and connecting the two URSP boards creates a testbed node with 4×4 MIMO transceiver configuration. The boards clock are synchronized, for ensuring synchronous transmission or reception, in such a way that the slave USRP-2953R board is exploiting the reference clock from the master USRP-2953R board.

The channel sounder application is based on Orthogonal Frequency Division Multiplexing (OFDM) modulation, where the reference sequences (pilots) are mapped onto the sub-carriers in a frequency interleaved fashion and transmitted by the multiple antenna ports. At the receiver side, the channel frequency response is computed over the positions where the reference sequences are mapped based on Least Square estimator, and then linearly interpolated in order to obtain a response across the entire transmission bandwidth. Furthermore, the reference sequences are generated using constant amplitude zero autocorrelation sequence (CAZAC); since the sequence has a constant amplitude property [10], which contributes a prominent advantage for good channel estimation.

Moreover, the multi-link MIMO channel sounder provides the possibility to sound carrier frequencies spanning from 1.2 to 6 GHz with 40 MHz of bandwidth. Given a network with N testbed nodes, the channel sounder application

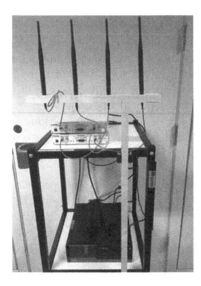

Fig. 1. Testbed node with 4×4 MIMO transceiver configuration.

allows estimating the $N \times N$ complex matrix of the channel responses between every couple of nodes within the network. The multi-link MIMO channel sounder application is based on the time division duplex (TDD) transmission mode; each node transmits his reference signals and receives the reference signals sent by the other nodes in a time interleaved fashion, as shown in Fig. 2. An estimate of the complex channel response for each link is obtained by demodulating the reference signals. The estimated complex channel response is sent to the testbed network controller through a backhaul network (Ethernet or WiFi) and logged for offline analysis.

		TDD frame														
N1		Tx	Rx	Rx	Rx	Rx	Rx	Rx	Rx	Rx	Rx	Rx	Rx		Tx
N2		Rx	Tx	Rx	Rx	Rx	Rx	Rx	Rx	Rx	Rx	Rx	Rx		Rx
N3		Rx	Rx	Tx	Rx	Rx	Rx	Rx	Rx	Rx	Rx	Rx	Rx		Rx
N4		Rx	Rx	Rx	Tx	Rx	Rx	Rx	Rx	Rx	Rx	Rx	Rx		Rx
.															
.															
.															
N12		Rx	Rx	Rx	Rx	Rx	Rx	Rx	Rx	Rx	Rx	Rx	Tx		Rx
		slot													

Fig. 2. The multi-link channel sounder with TDD transmission mode for 12 SDR nodes.

During the offline processing, the estimated complex channel matrices of desired and interfering signal are provided as an input to the Signal-to-Interference plus Noise (SINR) estimator based on selected receivers type using a hybrid simulation approach. This estimated SINR is mapped to throughput using the Shannon-Hartley formula, taking into account the highest order modulation being 256QAM (maximum spectral efficiency of 8 bits/s/Hz).

3   Measurement Campaign in Indoor Environment

In wireless networks, the reliability of the network performance studies notably depends on the employed scenarios and propagation models. For instance, in any simulation study, it is expected that the propagation model should be able to reproduce precisely the conditions of any link as it is in real network deployment. However, this condition will not be achieved all the time due to the presence of several indoor building irregularities and also ever change of building materials. The work in [11] investigated the WINNER II path loss model in an indoor environment, and their observations show that such model does not accurately predict the path loss on the selected links. In this paper, the evaluation employs real channel measurement data rather than relying on the statistic based propagation models as described in previous Sect. 1.

The channel measurement campaign has been executed in an indoor office scenario using our Multi-link MIMO channel sounder testbed. The measurement campaign is carried out at 5 GHz band with a transmission bandwidth of 20 MHz, and a transmit power of 10 dBm per antenna. During the campaign, the channel measurements are collected over 20 carrier frequencies spanning from 4.9 GHz to 5.8 GHz in order to experience different propagation conditions. Furthermore, multiple deployments with different propagation and geometrical characteristics have been taken into account to create different interference characteristics.

The channel measurement is attained in a typical indoor office environment with two pairs of adjacent rooms separated by a corridor as shown in Fig. 3, located at Aalborg University premises. As the figure depicts, the rooms have different size and divided by a concrete and plaster wall. These characteristics affect the signal propagation. In addition, there are also office furniture, black/white boards, and large office table lamps in the rooms which introduce a significant impact on the signal propagation pattern. Furthermore, the widespread presence of clutters and different building geometries plays a major role in shaping the interference characteristics.

The indoor office channel measurement campaign considered 18 possible different spatial locations as shown in Fig. 3. The campaign is carried out using 12 testbed nodes with 6 different deployments conditions for emulating different interference levels. For instance, one possible deployment of testbed node locations is highlighted with red dot circle in Fig. 3.

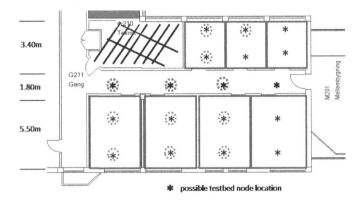

Fig. 3. Indoor office scenario - 18 testbed node location. (Color figure online)

4   Performance Evaluation

In this section, we present the network throughput performance results of our experimental analysis. The throughput results are computed for both an inter-ference suppression receiver such as IRC, and an interference-unaware MIMO receiver, known as Maximum Ratio Combining (MRC), which exploits the MIMO antenna cardinality to maximize the power of the desired signal.

The performance results are generated using hybrid simulation approach con-sidering 6 User Equipment (UEs), with Open Subscriber Group (OSG) access mode. The AP selection is based on the received power signal strength; each UE is connected to the AP which provides the highest received power. If there is an AP which does not serve any UE, it will be switched off and if an AP serves more than one UE, the UEs will equally share the available frequency resources.

The network performance is illustrated by the Empirical Cumulative Distri-bution Functions (ECDF) of users downlink throughput. In addition, three key performance indicator, namely Outage (5%-tile), median (50%) and peak (95%-tile), which are extracted from the ECDF results, are used in our performance analysis. Furthermore, during offline analysis, the throughput is scaled from transmission bandwidth of 20 MHz to bandwidth of our envisioned 5G concept (200 MHz) [12].

Figure 4 shows the ECDF of UEs throughput for different MIMO transceiver antenna configurations, considering a single stream transmission. The perfor-mance results are generated for both IRC and interference-unaware receivers considering 6 cells with the OSG access mode. The IRC receiver boosts the throughput performance with respect to the interference-unaware receiver for both MIMO configurations in such dense small cells network. The outage data rate gain of IRC receiver over the interference-unaware receiver is around 45% for a 2×2 MIMO configuration and 107% for a 4×4 MIMO configuration. The IRC receiver with single stream transmission mode uses the extra degree of freedom to suppress one interfering stream in 2×2 MIMO configuration cases and suppress up to three interfering streams in a 4×4 MIMO configuration case.

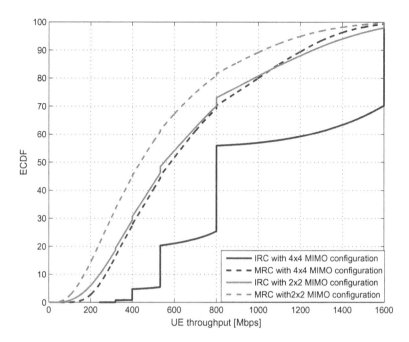

Fig. 4. The ECDF of users throughput with different MIMO transceivers configuration, considering six deployed cells using both IRC and MRC receivers.

The higher order MIMO antenna configuration improves the throughput performance in such uncoordinated dense small cells network for both receivers as demonstrated by extracted KPIs in Table 1; this is conventionally due to the gain on desired signal strength, and the IRC equipped system exploits the MIMO antenna cardinality for interference resilience purpose rather for spatial multiplexing. The results indicate that such deployments which lead to interference challenging situation are benefiting from a higher MIMO antenna configuration. Furthermore, the outage data rate gain of higher order MIMO configuration with respect to a lower MIMO configuration is significant when the users employ IRC receiver. Such significant gain is indeed achieved by the interference suppression capability of IRC receiver by suppressing three interferer cells 70% of the time in the case of 4×4 MIMO configuration with a single stream transmission mode. Since the number of active cells are less than or equal to four for 70% of the time as shown in Fig. 5, which means the users will face a maximum of three interfering streams with a single stream transmission mode.

Figure 6 illustrates the ECDF of the user's throughput for a different number of deployed cells, considering both interference suppression and interference-unaware MIMO receivers with a single stream transmission. The users throughput performance is improved with increasing number of deployed cells when the system employs IRC receiver. In the contrary, this does not hold for the interference-unaware system. Further, when the number of deployed cells is

Table 1. The data rate performance in Mbps for both 2×2 and 4×4 MIMO antenna configurations for both receiver types (MRC and IRC), considering six deployed cells.

	Outage		
	2×2 MIMO configuration	4×4 MIMO configuration	Gain (4×4 over 2×2)
IRC	187.6	470.3	150.7%
MRC	129.5	227.9	75.9%
	Median		
	2×2 MIMO configuration	4×4 MIMO configuration	Gain (4×4 over 2×2)
IRC	550.4	800.0	45.7%
MRC	440.2	581.6	32.1%
	Peak		
	2×2 MIMO configuration	4×4 MIMO configuration	Gain (4×4 over 2×2)
IRC	1443.0	1600.0	10.9%
MRC	1205.0	1353.0	12.3%

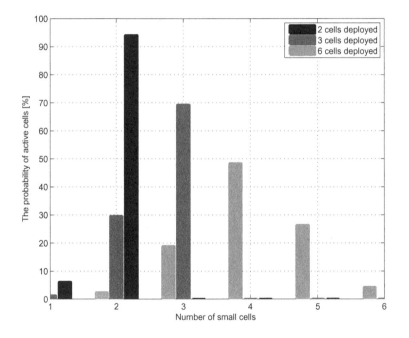

Fig. 5. The probability of active cells with respect to a number of deployed cells.

increasing, the coverage of the network will increase and the number of users sharing the available resources per cell will decrease since the users will select other cells which provide them strong signal strength. This is illustrated with the probability of active cells, which are selected by the six users, as depicted in Fig. 5. On the other hand, the interference conditions are becoming more

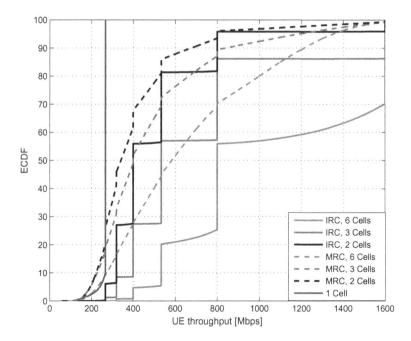

Fig. 6. The ECDF of users throughput with different number of deployed cells using both IRC and MRC receivers - the APs and the UEs are equipped with 4×4 MIMO transceivers.

challenging and introduce a limitation on the throughput performance. However, interference suppression receiver equipped users are coping with increasing number of uncoordinated cells. This is also illustrated in Table 2 in terms of outage, median, and peak data rate gain with respect to one cell deployment. The peak and median throughput gains are significant for a higher number of cells for both receivers. Increasing the number of deployed cells induces a loss in the outage data rate when the system employs inter-cell interference-unaware receiver. For instance, when the number of deployed cells is increased to three, the user's outage data rate gain is severely compromised with the interference-unaware receiver. However, the IRC receiver provides up to $+33\%$ outage data gain with respect to having one deployed cell, by using the extra degree of freedom to suppress the two interfering cells. Therefore, one can claim that uncoordinated dense small cell deployments will significantly suffer from interference if the system utilizes interference-unaware receivers; increasing the number of cells will not alleviate the performance of vulnerable users substantially. On the contrary, increasing the number of uncoordinated cells can alleviate the performance if the system applies advanced receiver processing with a sufficient MIMO antenna configuration.

Table 3 presents the outage data rate of a different number of cells when the users are equipped with a 2×2 MIMO configuration. The IRC throughput

Table 2. The data rate performance gain of increasing number of deployed cells with respect to one cell for both receiver types (MRC and IRC), the nodes are equipped with a 4×4 MIMO configuration.

	Outage		
	Gain 2 cells over 1 cell	Gain 3 cells over 1 cell	Gain 6 cells over 1 cell)
IRC	11.0%	33.2%	95.8%
MRC	−19.5%	−19.3%	−5.3%
	Median		
	Gain 2 cells over 1 cell	Gain 3 cells over 1 cell	Gain 6 cells over 1 cell)
IRC	50.0%	100.0%	200.0%
MRC	28.1%	50.0%	118.6%
	Peak		
	Gain 2 cells over 1 cell	Gain 3 cells over 1 cell	Gain 6 cells over 1 cell)
IRC	200.0%	500.0%	500.0%
MRC	200.0%	345.4%	406.9%

performance improvement with increasing number of cells does not hold. This is mainly because the IRC receiver has only one extra degree of freedom to suppress a strongest interferer stream in a 2×2 MIMO configuration equipped system with a single transmission mode. In addition, the IRC receiver throughput performance improvement will be challenged due to the lack of dominant interferer when the network consists of multiple cells with lower interfering levels. This has been noticed for a higher number of deployed cells. Generally, the operator can employ uncoordinated small cells deployment (which requires less planning) and rely on the advanced receiver with an acceptable MIMO antenna configuration to alleviate the interference burden.

Table 3. The outage data rate in Mbps for different number of deployed cells considering both receiver types (MRC and IRC), the nodes are equipped with a 2×2 MIMO configuration.

	1 cell	2 cells	3 cells	6 cells
IRC	207.6	**266.7**	196.7	187.6
MRC	207.6	116.9	112.4	129.1

5   Conclusions and Future Work

In this paper, we demonstrated the usage of interference suppression receivers in a practical dense small cells network with a higher MIMO antenna configuration using a hybrid simulation evaluation approach where propagation data are obtained from experimental analysis. The experimental study has been carried out in a typical indoor office scenario, using a multi-link MIMO channel sounder testbed with 12 testbed nodes which can be configured as AP or UE, where each node can also feature a 4×4 or a 2×2 MIMO configurations. The testbed is developed using USRP-2953R hardware and LabVIEW communication suite software package by National Instruments. The evaluation results prove that the interference suppression receiver improves the throughput performance with respect to the interference-unaware MIMO receiver for both a 2×2 and a 4×4 MIMO configuration setup. The outage data rate gain of IRC receiver over the interference-unaware receiver is around 45% for a 2×2 MIMO configuration and 107% for a 4×4 MIMO configurations. Advanced receiver processing with a higher order MIMO antenna configuration boosts the throughput performance significantly in practical dense small cells deployment. With respect to that, the IRC receiver with 4×4 MIMO antenna configuration achieves +150% outage throughput performance gain with respect to 2×2 MIMO configuration. The evaluation results also show that increasing the number of cells will not alleviate the throughput performance when the system utilize interference-unaware receiver. On the contrary, increasing the number of uncoordinated cells can alleviate the performance if the system applies advanced receiver processing with a sufficient MIMO antenna configuration. Furthermore, the results indicate that the operator can deploy the dense small cells uncoordinated and the interference can be tackled by advanced receiver processing with sufficiently capable MIMO antenna configuration.

Our future work will focus on the experimental investigation on the potential benefit of uncoordinated dense small cell deployment with interference suppression receivers capability with respect to distributed antenna system in an indoor office environment.

References

1. Nokia Solutions and Networks: Ten key rules of 5G deployment enabling 1 Tbit/s/km2 in 2030. Nokia Networks white paper (2015)
2. IMT vision framework and overall objectives of the future development of IMT for 2020 and beyond. International Telecommunication Union (ITU), Radio Communication Study Groups (2015)
3. Bhushan, N., et al.: Network densification: the dominant theme for wireless evolution into 5G. IEEE Commun. Mag. **52**(2), 82–89 (2014)
4. Mogensen, P., et al.: Centimeter-wave concept small cells. In: IEEE 79th Vehicular Technology Conference (2014)
5. Choi, J.: Optimal Combining and Detection Statistical Signal Processing for Communications. Cambridge University Press, Camberidge (2010)
6. Tavares, F.M.L., Berardinelli, G., Mahmood, N.H., Sørensen, T.B., Mogensen, P.: On the potential of interference rejection combining in B4G networks. In: IEEE 78th Vehicular Technology Conference, VTC2013-Fall (2013)
7. Wassie, D.A., Berardinelli, G., Tavares, F.M.L., Sørensen, T.B., Mogensen, P.: Experimental verification of interference mitigation techniques for 5G small cells. In: IEEE 81st Vehicular Technology Conference, VTC2015-Spring (2015)

8. National Instrument (2016). http://www.ni.com/labview-communications
9. National Instrument (2016). http://www.ni.com
10. Wen, Y., Huang, W., Zhang, Z.: CAZAC sequence and its application in LTE random acess. In: IEEE Information Theory Workshop (ITW) (2006)
11. Tonelli, O.: Experimental analysis and proof-of-concept of distributed mechnisms for local area wireless network. Ph.D. thesis (2014)
12. Mogensen, P., et al.: 5G small cell optimized radio design. In: IEEE Globecom, International Workshop on Emerging Techniques for LTE-Advanced and Beyond 4G (2013)
13. Lampinen, M., Carpio, F.D., Kuosmanen, T., Koivisto, T., Enescu, M.: System-level modeling and evaluation of interference suppression receivers in LTE system. In: IEEE 78th Vehicular Technology Conference, VTC2013-Fall (2013)

Generalized Minimum Distance Decoder in a DHA FH OFDMA Employing Concatenated Coding

Alexander Subbotin[1($\boxtimes$)] and Dmitry Osipov[1,2]

[1] National Research University Higher School of Economics,
3 Kochnovsky Proezd, Moscow 125319, Russia
`subbotin123@ya.ru`
[2] Institute for Information Transmission Problems,
Russian Academy of Sciences, 19 Bolshoy Karetny Lane, Moscow 127994, Russia

Abstract. In what follows a coded DHA FH OFDMA employing robust reception and concatenated coding construction is considered. The problem of employing Generalized Minimum Distance (GMD) decoder as an outer code decoder is considered. The effectiveness of the proposed decision is verified by means of simulation. In particular, performance of the communication system under intensive mixed interference is considered.

Keywords: Multiple access · Coded DHA FH OFDMA · Robust reception · Generalized Minimum Distance decoder · Iterative decoding

1 Introduction

Interference mitigation is one of the key issues in modern telecommunication systems design (see e.g.). This is mainly due to the fact that interference can be caused by different factors: authorized users' activity in a multiple access system (multi-user interference, MUI), signals transmitted by the users of other telecommunication systems operating within the same frequency bands or intentional jamming. If the interference is severe traditional reception techniques turn out to be ineffective due to low reliability of the computed decision statistics. Recently several coded modulation schemes [1–5] employing robust reception techniques were proposed to solve the problem. Due to their immunity to different types of interference they can be thought of as promising candidates for certain communication systems (e.g. M2M communications). In what follows a coded DHA FH OFDMA employing robust reception and concatenated coding construction is considered. In particular, decoding of the outer code by means of the Generalized Minimum Distance (GMD) decoder will be considered.

This paper is organized as follows. In Sect. 2 a short description of a coded DHA FH OFDMA system is given. In Sect. 3 a description of two different types

---

The results in Sect. 7 were obtained by Dmitry Osipov at the IITP RAS and financed by the Russian Science Foundation grant (project No. 14-50-00150).

T.K. Madsen et al. (Eds.): MACOM 2016, LNCS 10121, pp. 15–29, 2016.
DOI: 10.1007/978-3-319-51376-8_2

of noncoherent detectors will be given. In Sect. 4 concatenated construction in use will be described and the criterion for reliability values estimation for inner codes decoding will be proposed. In Sect. 5 the simulation scenario under consideration will be discussed. The effectiveness of the proposed approach will be verified by means of simulation in Sects. 6 and 7 respectively. Finally in Sect. 8 the obtained results will be summarized.

2   A DHA FH OFDMA System: Transmission and Reception

Let us consider a multiple access system in which K active users transmit information via a channel split into Q identical nonoverlapping subchannels by means of OFDM. In what follows it will be assumed that information that is to be transmitted is encoded into a codeword of a q-ary (n, k, d) block code $(q < Q)$. This code will be further on referred to as "inner code in time domain" (in what follows a specific case will be considered. In particular we shall consider the case in which the inner code in time domain is a short Maximum Distance Separable (MDS) code obtained by puncturing form a longer Reed-Solomon (RS) code). Whenever a user is to transmit a $q-$ary symbol it places 1 in the position of the vector $\bar{a}_g$ corresponding to the symbol in question within the scope of the mapping in use (in what follows it will be assumed that all the positions of the vector are enumerated from 1 to Q, moreover for the sake of simplicity and without loss of generality we shall assume that the 1st subchannel corresponds to 0, the 2nd subchannel corresponds to 1 and so on). Thus, each $q - ary$ symbol to be transmitted is mapped in to a weight 1 binary vector which will be referred to as "inner code in frequency domain". Then a random permutation of the aforesaid vector is performed and the resulting vector $\pi_g(\bar{a}_g)$ is used to form an OFDM symbol (permutations are selected equiprobably from the set of all possible permutations and the choice is performed whenever a symbol is to be transmitted). The transmission technique in question can be interpreted in the following way: assume that whenever a certain user is to transmit a symbol it randomly chooses q out of Q available subchannels (subcarriers). Since the list of the subchannels that can be used to transmit a signal (or a hopset) is allocated to each user in a dynamic fashion the technique under consideration is referred to as Dynamic Hopset Allocation Frequency Hopping OFDMA (DHA FH OFDMA). An example of a transmitter employing the strategy considered above is shown in Fig. 1 (symbols corresponding to the signals transmitted by the user under consideration are shown in italics). Therefore in order to transmit a codeword a user is to transmit n OFDM symbols. A sequence of OFDM symbols corresponding to a certain codeword that has been sent by a certain user will be referred to as a frame. Note that frames transmitted by different users need not be block synchronized, i.e. if within the time interval a certain user transmits a frame that corresponds to a codeword, symbols transmitted by another user within the same time period do not necessarily all comprise one codeword. Moreover, it will be assumed that transmissions from different

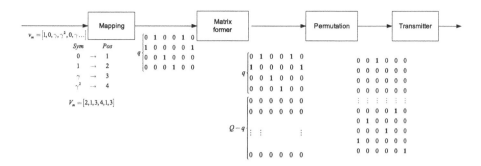

Fig. 1. Transmitter structure (for the case $q = 4$)

users are uncoordinated, i.e. none of the users has information about the others. In what follows we shall assume that all users transmit information in OFDM frames and the transmission is quasisynchronous. In terms of the model under consideration this assumption means that transmissions from different users are symbol synchronized.

Within the scope of a certain codeword reception the receiver is to receive n OFDM symbols corresponding to the codeword in question. Note that the receiver is assumed to be synchronized with transmitters of all users. Therefore all the permutations performed within the scope of transmission of the codeword in question are known to the user. The receiver measures energies at the outputs of all subchannels (let us designate the vector of the measurements as b_g where g is the number of the OFDM symbol) and applies inverse permutation to each vector b_g corresponding to the respective OFDM symbol thus reconstructing the initial order of elements and obtaining vector $\tilde{b}_g = \pi_g^{-1}(b_g)$. Let us consider a matrix X that consists of vectors $\tilde{b}_g = \pi_g^{-1}(b_g)$ that correspond to the codeword of inner code. Let us consider the submatrix $\Re = [\bar{\Re}_1, \bar{\Re}_2, \ldots, \bar{\Re}_n]$ (here $\Re$ is submatrix corresponding to the q first rows of the matrix X and each vector $\bar{\Re}_s$ is the height q column vector corresponding to the $s-th$ symbol of the codeword). Please note that $\Re$ provides all the information necessary to decode the codeword of the inner code.

3   Detection

The detection procedure that will be considered in this section can be decomposed in two successive procedures: reliability values computation and decoding. First and foremost let us consider the former procedure.

Let us assume that within the scope of the detection technique in use a reliability value is computed for each element of the matrix $\Re$. Let us designate the matrix of reliability values corresponding to the matrix $\Re$ with $Y = [\bar{Y}_1, \bar{Y}_2, \ldots, \bar{Y}_n]$. The main idea of the detection techniques under consideration is to use the reliability values stored in the matrix Y to compute decision statistics (reliability values) for codewords of the inner code in time domain.

Thus for the sake of brevity the matrix Y will be further on referred to as "decision matrix". The detector considered in [1] computes decision statistics called ranks. The easiest way to introduce ranks is to consider the indicator function of the following form:

$$I(x^*, x) = \begin{cases} 1 & x \le x^* \\ 0 & x > x^* \end{cases} \tag{1}$$

Rank of the element of the matrix $\Re(i, j)$ is given by:

$$\rho(i, j) = \sum_{k \ne i} \sum_{m \ne j} I\left(\Re(i, j), \Re(k, m)\right). \tag{2}$$

In other words a certain element $\Re(i, j)$ has rank ϱ if there are ϱ elements in the matrix having value lower than that of the element $\Re(i, j)$.

In RANK decoder introduced in [1] the decision matrix is the rank matrix considered above. The structure of the communication system employing the α detector is depicted in Fig. 2 (symbols corresponding to the signals transmitted by the user under consideration are shown in bold).

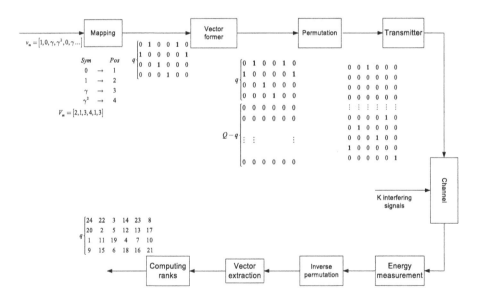

Fig. 2. A communication system employing RANK detector (for the case $q = 4$)

The RANK detector considered above requires rank matrix calculation. The latter operation requires sort of the elements of the $q \times n$ size matrix (where n is the codeword length of the inner code in time domain). Moreover in the case under consideration the decoding delay is lower bounded by the frame duration. In [5] a modified reception technique that provides lower complexity and delay has been proposed (in what follows it will be referred to as α detector). Let us assume that each column of the matrix $\Re$ is sorted in the descending order. Let us designate the α-th element of the vector v_j obtained by sorting the $j - th$ column of the matrix $\Re$ in the descending order by $v_j(\alpha)$. Let us consider the matrix D^α:

$$D^\alpha (i, j) = \begin{cases} 1 & \Re (i, j) \geq v_j (\alpha) \\ 0 & \Re (i, j) < v_j (\alpha). \end{cases} \qquad (3)$$

Each column of the matrix D^α contains exactly α nonzero entries. The nonzero entries in a certain column correspond to the elements of the respective column of the matrix $\Re$ having values greater or equal than α-th q quantile of this column. The matrix D^α is the decision matrix for the detector under consideration.

In Fig. 3 the structure of the communication system employing α detector is depicted (symbols corresponding to the signals transmitted by the user under consideration are shown in bold, symbols corresponding to the nonzero elements of the decision matrix are shown in bold italics).

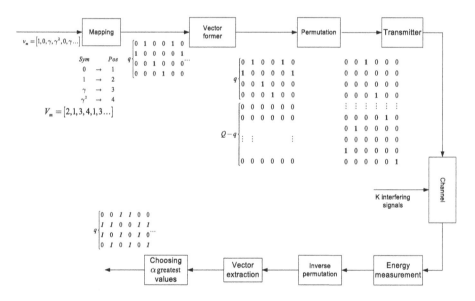

Fig. 3. A communication system employing α detector (for the case $q = 4$, $\alpha = 2$)

Let us assume that the user under consideration transmits a codeword v_m and within the scope of the mapping in use this vector is mapped into a matrix $X^m = [\bar{X}_1, \bar{X}_2, \ldots, \bar{X}_n]$. Let us denote the matrix

$$M = X^m * Y \qquad (4)$$

where Y is the decision matrix (i.e. $Y = \rho$ for the RANK detector and $Y = D^{\alpha}$ for the α detector) and $*$ stands for Hadamard product. The decision statistic for the mth codeword is then given by:

$$S_m = \sum_{i=1}^{n} \sum_{j=1}^{q} M(i, j) \tag{5}$$

The decoding rule boils down to choosing codeword number $m^* = \arg\max_{m} (S_m)$ if m^* is unique (i.e. $\forall\, m = 1 : \Omega, m \neq m^*\ m^* > m$, where $\Omega = |C|$), otherwise a denial decision is taken.

4   Concatenated Code and Generalized Minimum Distance Decoder

Even though it has been shown in [1–3,5] that the detectors described in the preceding section can provide relatively low probabilities of erroneous decoding and denial even under drastic interference the probabilities in question are not sufficiently low to meet the QoS requirements of modern communication systems. Thus some kind of concatenated coding is to be used. In what follows we shall consider the simplest case: the information to be transmitted is encoded with an outer (N, K, D) error correcting code over $GF(q^k)$ each symbol of this code being treated as k tuple over $GF(q)$ which in turn is encoded with an (n, k, d). Within the scope of the conventional decoding algorithm decision taken by the inner code decoder (in the case under consideration an inner decoder is replaced by a detector) are transferred directly to the decoder of the outer code i.e. in case of successful decoding (correct or erroneous) the symbol of the outer code corresponding to the respective codeword of the inner code is chosen, whereas denial decision results in erasure of the respective symbol of the outer code. Thus the decoder of the outer code is to correct both errors and erasures. If the number of errors and erasures is beyond the error correction capability of the outer code in use decoding can result in denial decision.

In this paper another approach that uses a classical Generalized Minimum Distance (GMD) decoder [6] is proposed. A GMD decoder makes use of symbol reliability values and combines it with the hard decision decoding. The codeword is decoded using a conventional decoding algorithm. If decoding fails (i.e. results in denial decision) two symbols with lowest reliability values are erased, the obtained vector being than decoded. The process is to be repeated iteratively until either the number of erasures to be corrected is beyond the error correction probability of the code in use or decoding is successful.

Within the scope of the reception strategy considered hereinabove reliability values for the symbols of outer code are not available. This is mainly due to the fact that channel state information cannot be evaluated with acceptable precision and thus the distributions of decision statistics cannot be computed. Thus the idea behind the proposed approach is to use information obtained from inner code decoder to evaluate the reliability of the decision taken by the inner code decoder. Let us consider a vector of decision statistics $\bar{S}^t = [S_1^t, S_2^t \ldots, S_M^t]$ corresponding to the t-th symbol of the outer code. Let us consider the vector $\bar{\Omega}^t = [\omega_1^t, \omega_2^t, \ldots, \omega_M^t]$ obtained by sorting the vector $\bar{S}^t$ in the descending order. Let us denote the reliability value λ^t for t-th symbol of the outer code as

$$\lambda^t = \begin{cases} \omega_1^t - \omega_2^t & \omega_1^t - \omega_2^t > 0 \\ n & \omega_1^t - \omega_2^t = 0 \end{cases} \tag{6}$$

The reliability value is computed for each matrix Y^t corresponding to the t-th symbol of the outer codeword. Thus the GMD decoding algorithm obtains the vector $\bar{r}$ of decisions obtained from inner decoder and the vector $\Lambda = [\lambda^1, \lambda^2, \ldots, \lambda^N]$ of reliability values (the last equality in Eq. 6 is to ensure that the symbols for which the erasure decision has been taken will not be erased once again by the GMD algorithm).

5   Simulation

To verify the effectiveness of the proposed solution simulation will be used. An OFDM system with Q subcarriers has been considered. Please note that no power control has been considered within the system under consideration i.e. the signals from the interfering users (at the receiver end) have powers κ greater than that of the signal from the user under consideration (in what follows we shall assume that $Q = 4096$ and $\kappa = 10^4$). The number of signals transmitted by interfering users is equal to K. It has been assumed that the received signal is affected by the broadband noise jamming. The power at the receiver end is described in terms of the signal to interference ratio SNR per bit (please note that the power of the broadband noise is determined in the entire band whereas the power of the signal is determined in the instantaneous band.

Since decoding of inner code boils down to exhaustive search the size of codebook should be moderate. Since the rate of inner code predetermines the overall concatenated code rate the length is to be relatively small. Thus in what follows we shall consider short block code as inner code in time domain. Hereinafter (n, k) MDS code obtained by puncturing $(q - 1, k, q - k)$ RS code will be used as inner code. We shall also choose (N, K) MDS code obtained by puncturing RS code over $GF(q^k)$. In particular, we shall assume inner codes with $k = 2$ over $GF(16)$ and $240, 160$ punctured RS code as outer code.

6   Simulation Results: GMD Decoding

In Figs. 4 and 5 the dependencies of the probability of undetected error (i.e. the probability of the fact that decoding will result in denial or erroneous decoding) for RANK detector, different rates of inner code and $K = 300$ are shown.

As can be seen from the obtained curves the proposed solution can decrease the probability of undetected error up to one order magnitude. The following curves (Figs. 6 and 7) show that the same effect can be archived for RANK detector for other values of K.

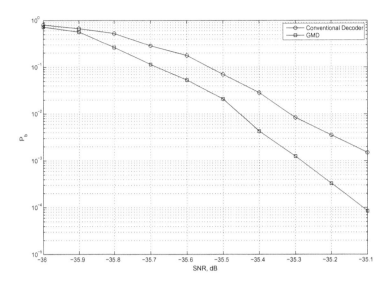

Fig. 4. Dependencies of the probability of undetected error on the value of SNR for RANK detector (number of interfering signals K = 300, inner code rate R = 1/3)

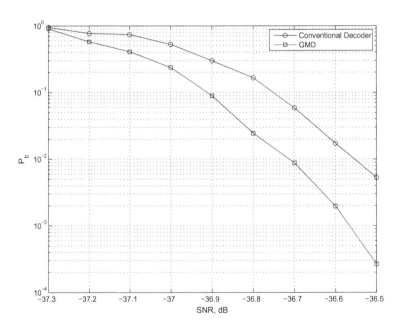

Fig. 5. Dependencies of the probability of undetected error on the value of SNR for RANK detector (number of interfering signals K = 300, inner code rate R = 1/4)

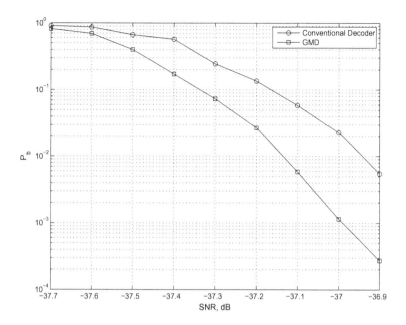

Fig. 6. Dependencies of the probability of undetected error on the value of SNR for RANK detector (number of interfering signals K = 200, inner code rate R = 1/4)

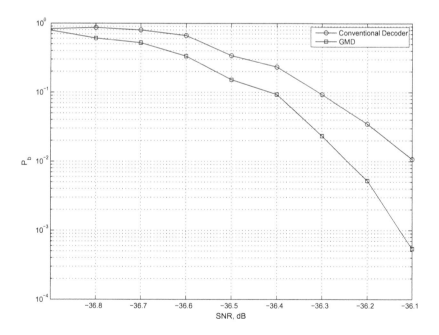

Fig. 7. Dependencies of the probability of undetected error on the value of SNR for RANK detector (number of interfering signals K = 400, inner code rate R = 1/4)

Since the outer code has sufficient error-correction capability the probability of denial is sufficiently higher than the probability of erroneous decoding. Thus it seems reasonable to consider the probability of denial separately. The respective dependencies are depicted in Fig. 8.

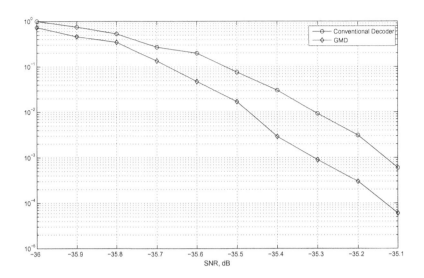

Fig. 8. Dependencies of the probability of denial on the value of SNR for RANK detector (number of interfering signals K = 300, inner code rate R = 1/5)

Let us now consider how the application of the purposed approach to GMD decoding influences the performance of the system employing α detector as inner decoder. As an example let us consider the performance of such system for inner code with rate $R = 1/6$ and various values of K (see Figs. 9 and 10).

Similarly to the case of the RANK detector the application of GMD results in substantial performance gain. The obtained curves demonstrate that the proposed solution can provide low probability of undetected error (per codeword) even under drastic interference. Finally, let us consider the effect of applying GMD on the denial probability of the outer code decoding see Fig. 11.

Again, similarly to the case of RANK detector the application of GMD in the case were α detector results in probability of denial decision decrease up to one order of magnitude.

7   Simulation Results: Restricted GMD Decoding

Hereinabove it has been demonstrated that by applying GMD as an outer code decoder in the way proposed in the previous sections the performance of the system can be significantly improved (as compared to that ensured by the conventional algebraic decoder correcting errors and erasure). However, this gain is

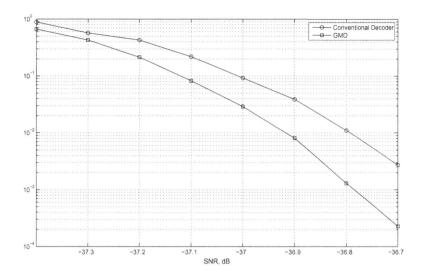

Fig. 9. Dependencies of the probability of undetected error on the value of SNR for α detector (number of interfering signals K = 400, inner code rate R = 1/6)

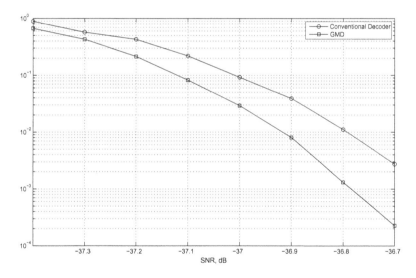

Fig. 10. Dependencies of the probability of undetected error on the value of SNR for α detector (number of interfering signals K = 600, inner code rate R = 1/6)

obtained at the expense of increased complexity and delay (due to the iterative nature of the decoding algorithm). Thus it seems interesting to investigate the case when the decoder is allowed to make only moderate number of iterations (i.e. not only is the number of symbols of the outer code to be erased is limited by the error correction capability of the code in use but rather is restricted in

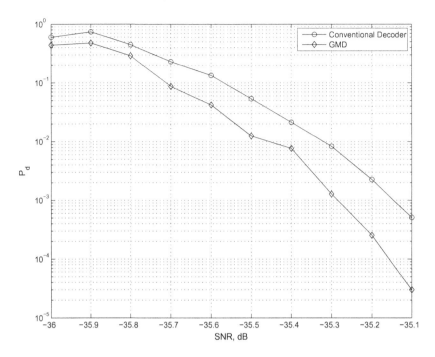

Fig. 11. Dependencies of the probability of denial on the value of SNR for α detector (number of interfering signals K = 500, inner code rate R = 1/5)

advance). In Fig. 12 the dependencies of the probability of undetected error on the value of SNR for the case when the number of iterations is upper bounded for the case K = 300 are depicted (the curves for the conventional decoder and for the unrestricted GMD decoder are depicted for comparison).

It can be noted that although the GMD decoder with limited number of iterations cannot provide the reliability that can be obtained with the unrestricted GMD decoder it is still possible to decrease the probability of undetected error substantially while with fixed complexity and delay. For instance for SNR values less than −35.15 dB GMD detector with 25 iterations can decrease the probability of undetected error (per codeword) by an order of magnitude or more. Let us now consider the denial probability. In Fig. 13 the dependencies of the denial probability on the value of SNR for different values of the maximum number of iterations allowed for the case K = 300 are depicted (the curves for the conventional decoder and for the unrestricted GMD decoder are depicted for comparison). As can be seen Figs. 13 and 12 almost coincide i.e. the probability of the fact that an uncorrectable error pattern will result in erroneous decoding is negligible as compared to the probability of the fact that an uncorrectable error pattern will result in denial.

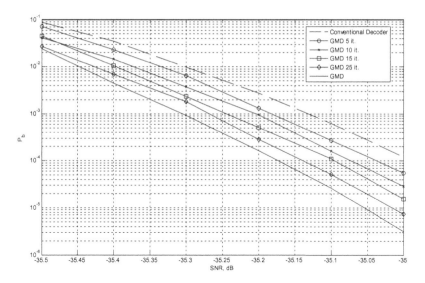

Fig. 12. Dependencies of the probability of undetected error (per codeword) on the value of SNR for RANK detector and various types of outer decoder (number of interfering signals K = 300, inner code rate R = 1/3)

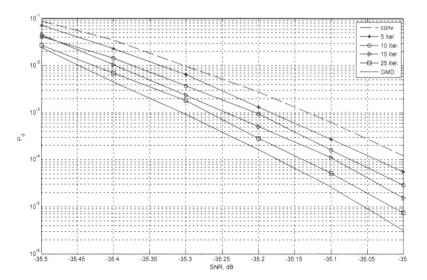

Fig. 13. Dependencies of the denial probability (per codeword) on the value of SNR for RANK detector and various types of outer decoder (number of interfering signals K = 300, inner code rate R = 1/3)

Let us now consider the case K = 200. In Fig. 14 the dependencies of the probability of undetected error (per codeword) on the value of SNR for this case and various values of maximal number of iterations are shown (the curves

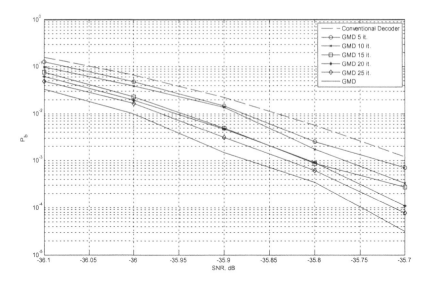

Fig. 14. Dependencies of the probability of undetected error (per codeword) on the value of SNR for RANK detector and various types of outer decoder (number of interfering signals K = 200, inner code rate R = 1/3)

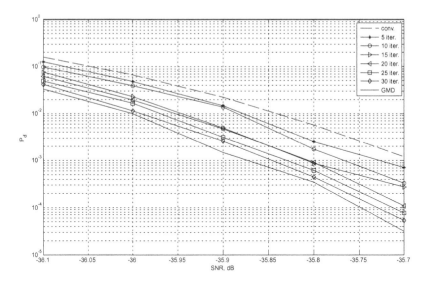

Fig. 15. Dependencies of the denial probability (per codeword) on the value of SNR for RANK detector and various types of outer decoder (number of interfering signals K = 200, inner code rate R = 1/3)

for the conventional decoder and for the unrestricted GMD decoder are also shown in Fig. 14 for comparison). The curves shown on Fig. 14 confirm that the probability of undetected error can be substantially reduced by applying

GMD decoder even if maximum number of iterations allowed is fixed i.e. the situation is very similar to that considered previously. Finally let us consider the dependencies of the denial probability on the value of SNR for different values of the maximum number of iterations allowed for the case $K = 200$. The respective curves are shown in Fig. 15.

Again we can claim that the curves for the denial probability and probability of undetected error almost coincide.

8 Conclusion and Future Work

In this paper a coded DHA FH OFDMA employing concatenated coding and a GMD as an outer code decoder have been considered. It has been demonstrated that the proposed approach can result in substantial reliability increase. It has been demonstrated that even if the number of iteration is upperbounded the GMD-type decoder proposed hereinabove can provide substantial reliability increase. Specific parameters choice is to be based on the tradeoff between the desired QoS and complexity and delay requirements. The choice in question essentially depends on the specific application and scenario. The applicability of the proposed approach to the specific communication scenarios is a subject of future work.

References

1. Kondrashov, K., Afanassiev, V.: Ordered statistics decoding for semi-orthogonal linear block codes over random non-Gaussian channels. In: Proceedings of the Thirteenth International Workshop on Algebraic and Combinatorial Coding Theory, Pomorie, Bulgaria, 15–21 June 2012, pp. 192–196 (2012)
2. Kreshchuk, A., Potapov, V.: New coded modulation for the frequency hoping OFDMA system. In: Proceedings of the Thirteenth International Workshop on Algebraic and Combinatorial Coding Theory, Pomorie, Bulgaria, 15–21 June 2012, pp. 192–196 (2012)
3. Osipov, D.: Inner convolutional codes and ordered statistics decoding in a multiple access system enabling wireless coexistence. In: Jonsson, M., Vinel, A., Bellalta, B., Marina, N., Dimitrova, D., Fiems, D. (eds.) MACOM 2013. LNCS, vol. 8310, pp. 33–38. Springer, Heidelberg (2013). doi:10.1007/978-3-319-03871-1_4
4. Osipov, D.: On jamming-proof signal-code constructions for a multiple access system. In: Proceedings of the XIV International Symposium on Problems of Redundancy in Information and Control Systems, 1–5 June 2014, pp. 74–77. IEEE Press (2014)
5. Osipov, D.: Reduced-complexity robust detector in a DHA FH OFDMA system under mixed interference. In: Jonsson, M., Vinel, A., Bellalta, B., Belyaev, E. (eds.) MACOM 2014. LNCS, vol. 8715, pp. 29–34. Springer, Heidelberg (2014). doi:10.1007/978-3-319-10262-7_3
6. Forney, G.D.: Generalized minimum distance decoding. IEEE Trans. Inf. Theory **IT−12**, 125–131 (1966)

Modemless Multiple Access Communications Over Powerlines for DC Microgrid Control

Marko Angjelichinoski$^{(\boxtimes)}$, Čedomir Stefanović, and Petar Popovski

Department of Electronic Systems, Aalborg University, Aalborg, Denmark
{maa,cs,petarp}@es.aau.dk

Abstract. We present a communication solution tailored specifically for DC microgrids (MGs) that exploits: (i) the communication potential residing in power electronic converters interfacing distributed generators to powerlines and (ii) the multiple access nature of the communication channel presented by powerlines. The communication is achieved by modulating the parameters of the primary control loop implemented by the converters, fostering execution of the upper layer control applications. We present the proposed solution in the context of the distributed optimal economic dispatch, where the generators periodically transmit information about their local generation capacity, and, simultaneously, using the properties of the multiple access channel, detect the aggregate generation capacity of the remote peers, with an aim of distributed computation of the optimal dispatch policy. We evaluate the potential of the proposed solution and illustrate its inherent trade-offs.

1 Introduction

MicroGrids (MGs) are localized clusters of small-scale Distributed Energy Resources (DERs) and loads that operate either connected to the main grid or in standalone mode [1,2]. The MG control plane is typically organized into primary, secondary and tertiary levels [2–4]. The primary control enables the basic operation of the system by regulating the electrical parameters (bus voltage and/or frequency) and keeping the supply-demand balance to guarantee stability. It is implemented in a decentralized manner using the *droop control law* [2,4,5], relying only on local measurements. The upper (secondary/tertiary) level control optimizes the performance of the MG in terms of maximizing the quality of the delivered power under minimal cost, and, in order to operate properly, requires exchange of information among DERs [6–8]. An important control application is the *Optimal Economic Dispatch (OED)*. OED runs periodically, e.g., every 5–30 min, and dispatches the DERs based on their generation capacities at minimum total cost. In MGs with predominantly stochastic renewable generation, the generation capacities of the DERs vary from one dispatch period to the next and they have to be reported regularly to the OED [6].

The traditional assumption is that an external communication system, such as wireless, covers the communication requirements of the upper level control [8]. However, recent works challenge this assumption due to the following issues:

© Springer International Publishing AG 2016
T.K. Madsen et al. (Eds.): MACOM 2016, LNCS 10121, pp. 30–44, 2016.
DOI: 10.1007/978-3-319-51376-8_3

(1) the external communication system may jeopardize the efficiency and stability of the MG due to limited reliability and availability [5], (2) the distributed power systems, particularly MGs, are significantly more dynamic, sporadic and ad-hoc in nature, compared to traditional centralized power system, which might deem the installation of external communication system impractical and cost inefficient [2,5,8–10], and (3) making the MG system reliant on an external system contradicts the principle of self-sustainability and self-sufficiency [12–14]. A suitable alternative is to use the existing power electronic and powerline equipment for communication [9–13]. *Power talk* is such solution, with a target use in direct current (DC) MGs [12–14]. Specifically, power talk modulates information into deviations of the parameters of the primary control loops of the DERs. In this way, a non-linear multiple access communication channel is induced, through which information-carrying deviations of the voltage (or, equivalently, power) are disseminated throughout the system, and received and processed by other DER units. The control frequency of the primary droop controller is typically between 10–1000 Hz, which implies that power talk is a narrowband solution. It exhibits similarities with other existing low-rate PLC standards for communication in the AC distribution grids, such as Ripple Carrier, TWACS and Turtle [15], which also rely on modulating voltage to exchange information. However, in contrast to these solutions, power talk requires no additional hardware, being implemented in the local primary control loop of the power electronic converters that connect the DERs to the DC buses. Thus, power talk fosters the self-sustainability feature of the MG paradigm.

Previous works focused on the communication-theoretic aspects of the power talk, including the design of robust communications under variable loads, which is the major communication impairment [12–14,16–18]. In this paper, we use power talk to support the upper level control optimizations. In particular, we focus on the OED and its distributed solution under linear incremental cost functions (i.e., cost per unit generation) [6]. We identify the information required by the DERs to run the dispatch in distributed manner, showing that it is sufficient to locally obtain the *aggregate* generation capacity of DERs with equal incremental cost. Based on this observation, we develop a communication and computation scheme which runs periodically, in dedicated time interval prior to each dispatch period. In the proposed scheme, DERs with equal incremental costs transmit quantized, uncoded information about their local capacities over the power talk channel in full duplex mode, whereas the receiving DERs directly detect the aggregate capacity of the transmitting DERs. The obtained information is then used to determine the optimal dispatch in distributed manner. The proposed solution can be viewed as a decentralized upper-level controller where the required communication capability is enabled by the primary control level that exploits the multiple-access nature of the powerlines interconnecting DERs.

The rest of the paper is organized as follows: Sect. 2 introduces the model of a DC MG and reviews the relevant aspects of the power talk multiple access channel. Section 3 briefly reviews the OED and discusses its distributed solution.

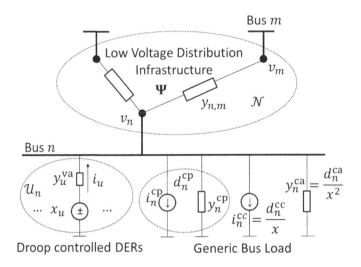

Fig. 1. System model: DC MicroGrid in steady state.

Section 4 presents the power talk based solution for distributed OED and Sect. 5 evaluates its performance. Section 6 concludes the paper.

2   Power Talk Multiple Access Channel

2.1   System Model

A DC MG is a collection of DERs and loads that are connected to Low Voltage DC (LVDC) distribution infrastructure, see Fig. 1. We denote the total number of DERs with U, indexed in the ordered set $\mathcal{U} = \{0, ..., U - 1\}$. The LVDC infrastructure consist of $N \geq 1$ buses, indexed in the ordered set $\mathcal{N} = \{0, ..., N - 1\}$; a *bus* is defined as a point in the MG characterized by a steady state voltage denoted with $v_n, n \in \mathcal{N}$. Bus $n \in \mathcal{N}$ hosts U_n DERs, where $0 < U_n < U$. We denote the corresponding subset of DERs by $\mathcal{U}_n$, where $|\mathcal{U}_n| = U_n$ and where $\mathcal{U}_n \subset \mathcal{U}$. We also introduce matrix $\mathbf{E} \in \{0, 1\}^{N \times U}$ with entries defined as:

$$e_{n,u} = \begin{cases} 1, & u \in \mathcal{U}_n, \\ 0, & \text{otherwise.} \end{cases} \tag{1}$$

The DERs use power electronic converters to interface the distribution infrastructure, and their voltage and current (i.e., power) outputs are locally regulated via primary control. A common primary control configuration is in the form of Voltage Source Converter (VSC), see Fig. 1, which regulates the output voltage and current using the following law [2,4,5]:

$$v_n = x_u - \frac{1}{y_u^{\text{va}}} i_u, \ u \in \mathcal{U}_n, n \in \mathcal{N}, \tag{2}$$

where i_u is the output current of the DER, x_u is the *reference voltage* and y_u^{va} is the *virtual admittance*. This implementation is known as decentralized droop control.[1] The droop controller controls x_u and y_u^{va}, where x_u determines the voltage rating of the system, while the y_u^{va} determines the power sharing among different DERs.[2] Besides DERs, bus $n \in \mathcal{N}$ also hosts a collection of loads, represented through an aggregate model comprising *constant admittance* $y_n^{ca} = x^{-2}d_n^{ca}$, *constant current* $i_n^{cc} = x^{-1}d_n^{cc}$, and *constant power* component d_n^{cp}. The quantities d_n^{ca}, d_n^{cc} and d_n^{cp} are the *rated power demands* of the individual components at specified voltage x. The buses are interconnected through distribution lines; the line connecting buses n and m is characterized by an *admittance*, denoted with $y_{n,m} \equiv y_{m,n} \geq 0$, with equality if buses n and m are not directly connected, or if $n = m$. The *admittance matrix* of the system is denoted with $\boldsymbol{\Psi} \in \mathbb{R}^{N \times N}$, with entries defined as:

$$\psi_{n,m} = \begin{cases} \sum_{i \in \mathcal{N}} y_{n,i}, & n = m, \\ -y_{n,m}, & n \neq m. \end{cases} \tag{3}$$

Applying Kirchoff's laws to the system depicted in Fig. 1 leads to the following current balance equation:

$$\sum_{u \in \mathcal{U}_n} (x_u - v_n) y_u^{va} = \sum_{m \in \mathcal{N}} (v_n - v_m) y_{n,m} + \frac{v_n}{x^2}d_n^{ca} + \frac{1}{x}d_n^{cc} + \frac{1}{v_n}d_n^{cp}, \; n \in \mathcal{N}. \tag{4}$$

Solving for v_n, $n \in \mathcal{N}$, yields the following expression [2,4]:

$$v_n = \frac{\sum_{u \in \mathcal{U}_n} x_u y_u^{va} + \sum_{m \in \mathcal{N}} v_m y_{n,m} - \frac{d_n^{cc}}{x}}{2 \left(\sum_{u \in \mathcal{U}_n} y_u^{va} + \sum_{m \in \mathcal{N}} y_{n,m} + \frac{d_n^{ca}}{x^2} \right)} +$$

$$\frac{\sqrt{\left(\sum_{u \in \mathcal{U}_n} x_u y_u^{va} + \sum_{m \in \mathcal{N}} v_m y_{n,m} - \frac{d_n^{cc}}{x} \right)^2 - 4d_n^{cp} \left(\sum_{u \in \mathcal{U}_n} y_u^{va} + \sum_{m \in \mathcal{N}} y_{n,m} + \frac{d_n^{ca}}{x^2} \right)}}{2 \left(\sum_{u \in \mathcal{U}_n} y_u^{va} + \sum_{m \in \mathcal{N}} y_{n,m} + \frac{d_n^{ca}}{x^2} \right)}.$$

$$\tag{5}$$

In power talk, each unit $u \in \mathcal{U}$ modulates information into the values of the local reference voltage and virtual admittance droop control parameters and observes the steady state bus voltage response [12,13,17]. In this regard, the

---

[1] Another primary control architecture is the Current Source Converter (CSC). CSC units do not participate in output voltage regulation and they are operated at their generation capacity regardless of the system state, using the Maximum Power Point Tracking (MPPT) algorithm. When the units engage in power talk communication, they are configured as VSC units, i.e., all CSC units, for the purpose of exchanging information via power talk switch to VSC mode of operation.

[2] In practice, the value of the virtual admittance is set to enable proportional power sharing. This aspect is discussed in more detail in Sect. 3.

model (5) describes the general input-output relation of the power talk *multiple access channel*, where the inputs are x_u and y_u^{va}, $u \in \mathcal{U}$, while the output observed by DER u, $u \in \mathcal{U}_n$, is v_n. We comment on several important aspects: (1) the obtained multiple access channel is non-linear due to the presence of constant power loads, (2) the output is determined by the physical configuration of the system, and (3) the output is determined by the current power demand of the load components. From (5), we observe that solving $v_n = v_n(x_0, ..., x_{U-1}, y_0^{\text{va}}, ..., y_{U-1}^{\text{va}})$, $\forall n \in \mathcal{N}$, requires knowledge of the admittance matrix and the power demands of the individual load components. On the primary control level, the DERs rely only on the output current as a feedback and they do not have knowledge of the MG configuration, making the power talk channel (5) very difficult to solve [17]. We tackle these difficulties by using a linearized approximation of the bus voltage around a predefined operating point, which does not require detailed knowledge of the admittance matrix and the load components [14].

2.2 Discrete Time Linearized Signal Model

We develop a linearized signal model for all-to-all full duplex power talk communication scenario, where all DERs simultaneously transmit and receive data. We assume that the time is slotted in slots of duration T_S and the units are slot-synchronized.[3] In slot t, DER u uses the following input:

$$x_u(t) = \overline{x}_u + \Delta x_u(t), \tag{6}$$
$$y_u^{\text{va}}(t) = \overline{y}_u^{\text{va}}, \; u \in \mathcal{U}, \tag{7}$$

with $\Delta x_u(t)$ being the *input* signal and $\overline{x}_u$, $\overline{y}_u^{\text{va}}$ are droop combinations defining the current operating point of the MG. In other words, in the considered power talk variant, the information is modulated into the deviations of the reference voltage droop control parameters while the virtual admittances remain fixed. The resulting deviation of the bus voltage in slot t can be written as:

$$v_n(t) = \overline{v}_n + \Delta v_n(t), \; n \in \mathcal{N}, \tag{8}$$

where $\Delta v_n(t)$ is the *output* of the communication channel, and where the steady state bus voltage $\overline{v}_n$ corresponds to $\overline{x}_0, ..., \overline{x}_{U-1}, \overline{y}_0^{\text{va}}, ..., \overline{y}_{U-1}^{\text{va}}$. Unit u samples the

---

[3] The duration of the time slot T_S is set to comply with the control frequency of the primary controller; its value is typically of the order of milliseconds to allow the system to establish steady-state. The aspects of achieving and maintaining slot-synchronized power talk operation is beyond the scope of the paper. However, we note that the power electronic converters may come pre-equipped with GPS modules, providing a common time reference for achieving and maintaining slot-synchronization. An alternative option is to use distributed algorithms for achieving synchronization, while standard line codes, e.g. Manchester code, can be used to maintain synchronization among the units.

noisy version of $\Delta v_n(t)$ with frequency f_S and uses the average of $N_S = T_S f_S$ samples over the slot t to obtain the observation:[4]

$$\Delta \tilde{v}_u(t) = \Delta v_n(t) + z_u(t), \ u \in \mathcal{U}_n, n \in \mathcal{N}, \tag{9}$$

where noise $z_u(t) \sim \mathcal{N}(0, \sigma^2)$ is modeled as additive Gaussian noise [17]. Finally, we assume that the loads in the system change randomly with a rate that is much lower than the signaling rate T_S^{-1} and that the signaling is done over a single realization of the load values.[5]

We assume that the reference voltage deviations $\Delta x_u(t)$ are relatively small:

$$\frac{|\Delta x_u(t)|}{\overline{x}_u} \ll 1 \Rightarrow \frac{|\Delta v_n(t)|}{\overline{v}_n} \ll 1, \ u \in \mathcal{U}_n, n \in \mathcal{N}. \tag{10}$$

Under this assumption, (8) can be linearized around $\overline{x}_0, ..., \overline{x}_{U-1}, \overline{y}_0^{va}, ..., \overline{y}_{U-1}^{va}$, yielding the following linear model [14]:

$$\Delta \mathbf{v}(t) \approx (\check{\boldsymbol{\Psi}} + \mathbf{K}^{-1}(\mathbf{E}\mathbf{Y}^{va}\mathbf{E}^T + \mathbf{Y}^{ca}))^{-1}\mathbf{E}\mathbf{Y}^{va}\Delta \mathbf{x}(t) = \check{\mathbf{H}}\Delta \mathbf{x}(t). \tag{11}$$

where $\Delta \mathbf{x}(t) = [\Delta x_u(t)]_{u \in \mathcal{U}}^T \in \mathbb{R}^{U \times 1}$, $\Delta \mathbf{v}(t) = [\Delta v_n(t)]_{n \in \mathcal{N}}^T \in \mathbb{R}^{N \times 1}$, $\check{\boldsymbol{\Psi}} \in \mathbb{R}^{N \times N}$ is the modified admittance matrix where each diagonal entry is multiplied by κ_n^{-1}, i.e. $\check{\psi}_{n,n} = \frac{\psi_{n,n}}{\kappa_n}$, $\mathbf{Y}^{va} = \mathrm{diag}\,\{y_u^{va}\}_{u=\mathcal{U}} \in \mathbb{R}^{U \times U}$, $\mathbf{Y}^{ca} = \mathrm{diag}\,\{y_n^{ca}\}_{n \in \mathcal{N}} \in \mathbb{R}^{N \times N}$, $\mathbf{K} = \mathrm{diag}\,\{\kappa_n\}_{n \in \mathcal{N}}$ and $\kappa_n \geq 1$ appears as a result of linearization (see [14]). We refer to matrix $\check{\mathbf{H}}$ as the *channel* matrix of the system. The obtained linear model for the noisy output observed by DER k, $k \in \mathcal{U}_n$, is:

$$\Delta \tilde{v}_k(t) \approx \sum_{m \in \mathcal{N}} \check{h}_{n,m} \sum_{l \in \mathcal{U}_m} \Delta x_l(t) + z_k(t), \tag{12}$$

$\check{h}_{n,m}$ is the entry at position n, m of $\check{\mathbf{H}}$; it can be shown that $\check{h}_{n,m} > 0, \forall n, m$.

Due to deviations of the reference voltages, the output power of the DERs will also deviate. Denote with $p_u(t) = v_n(t)i_u(t)$ the output power of DER $u \in \mathcal{U}_n$ in slot t. Using assumption (10), $p_u(t)$ can be approximated as:

$$p_u(t) = \overline{p}_u + \Delta p_u(t) \tag{13}$$

$$\approx \overline{p}_u + \sum_{m \in \mathcal{N}} \check{\phi}_{u,m} \sum_{l \in \mathcal{U}_m} \Delta x_l(t), \tag{14}$$

where $\overline{p}_u$ corresponds to $\overline{x}_0, ..., \overline{x}_{U-1}, \overline{y}_0^{va}, ..., \overline{y}_{U-1}^{va}$, and where:

$$\check{\phi}_{u,m} = \begin{cases} (\check{h}_{n,m}\overline{x}_u - 2\check{h}_{n,m}\overline{v}_n)\overline{y}_u^{va}, & m \neq n, \\ (\check{h}_{n,m}\overline{x}_u + (1 - 2\check{h}_{n,m})\overline{v}_n)\overline{y}_u^{va}, & m = n. \end{cases} \tag{15}$$

In practice, the amount of deviation of the output power that can be tolerated is a design parameter that constraints the input power talk signal $\Delta x_u(t)$.

---

[4] More precisely, the bus voltage is sampled after the system reaches a steady state and all transient effects diminish.

[5] Typically, the average time between consecutive load changes in MG systems is of the order of several seconds or even minutes [6,8].

3   Distributed Optimal Economic Dispatch

Here we briefly review the distributed OED [6–8]. From the perspective of the OED, the DERs are organized in G disjoint subsets/types. The subsets are denoted with $\mathcal{U}_g$, $g = 0, ..., G-1$. Each subset is assigned incremental cost c_g per unit of generated power, where the cost c_g is the same for all DERs in the subset. Without loss of generality, assume the costs are ordered as $c_0 \leq c_2 \leq ... \leq c_{G-1}$. We introduce the binary matrix $\Xi \in \{0,1\}^{G \times U}$, with entries defined as:

$$\xi_{g,u} = \begin{cases} 1, & u \in \mathcal{U}_g, \\ 0, & \text{otherwise.} \end{cases} \tag{16}$$

The generation capacity of DER u at the beginning of each dispatch period is denoted with w_u. The aggregate generation capacity of all DERs of the same type g is denoted with $w^{(g)} = \sum_{u \in \mathcal{U}_g} w_u$. The total power demand of all loads in the system during a single dispatch period is $d^L = \sum_{n \in \mathcal{N}} (d_n^{ca} + d_n^{cc} + d_n^{cp})$. We assume a typical demand-response scenario where the total load demand d^L is known *a priori* (e.g., through accurate forecast programs). In such case, the goal of the OED is to dispatch the available DER resources in optimal manner.

We define the following: (1) the *power generation capacity vector* $\mathbf{w} = [w_u]_{u \in \mathcal{U}} \in \mathbb{R}^{U \times 1}$, (2) the *generation cost vector* $\mathbf{c} = [c_g]_{g=0,...,G-1} \in \mathbb{R}^{G \times 1}$, and (3) the *dispatch policy vector* $\mathbf{p} = [p_u]_{u \in \mathcal{U}} \in \mathbb{R}^{U \times 1}$. The total power generation cost in a dispatch period is:

$$C(\mathbf{p}; \mathbf{w}, \mathbf{c}, d^L) = \sum_{g=0}^{G-1} \sum_{u \in \mathcal{U}_g} c_g p_u. \tag{17}$$

The *optimal dispatch policy* $\mathbf{p}^\star$ is solution to the optimization problem:

$$\mathbf{p}^\star = \min_{\mathbf{p}} \ C(\mathbf{p}; \mathbf{w}, \mathbf{c}, d^L) \tag{18}$$

$$\text{s.t.} \ \sum_{g=0}^{G-1} \sum_{u \in \mathcal{U}_g} p_u = d^L,$$

$$0 \leq p_u \leq w_u, u \in \mathcal{U}_g, g = 0, ..., G-1.$$

It can be shown that the following distributed policy is optimal for (18) [6]:

$$p_u^\star = \begin{cases} w_u, & d^L > \sum_{j=0}^{g} w^{(j)}, \\ 0, & d^L < \sum_{j=0}^{g-1} w^{(j)}, \\ w_u \frac{(d^L - \sum_{j=0}^{g-1} w^{(j)})}{w^{(g)}}, & \sum_{j=0}^{g-1} w^{(j)} \leq d^L \leq \sum_{j=0}^{g} w^{(j)}. \end{cases} \tag{19}$$

The first condition in (19) configures the DER as constant power source to inject the maximum available power into the system. The second condition sets the unit in idle mode, i.e., the DER does not inject power into the system. The third condition configures the DER as VSC unit for proportional power sharing, i.e., the DER employs droop control with virtual resistance set to enable proportional power sharing based on the rating w_u.

From (19) it can be noted that DERs of type $g = 0, ..., G - 1$, require the knowledge of the aggregate generation capacities $w^{(k)}$, $k \le g$, to make the local decision, while knowledge of $w^{(k)}$, $k > g$, is not necessary. Based on this observation, in the following section we design a power talk communication protocol to facilitate (19).

4   Power Talk for Distributed OED

4.1   Organization of the Protocol Operation

Typically, economic dispatch is run periodically, with a period T ranging from 5 up to 30 min [6]. We split this period into two phases, see Fig. 2: (1) *communication phase* of duration T_{PT}, in which the units exchange information via the power talk multiple access channel (12), and (2) *OED phase* of duration $T_{ED} = T - T_{PT}$, in which the configuration of each DER is determined by the outcome of the optimal decentralized algorithm (19), based on the information obtained during the communication phase.[6] The communication phase is split in G sub-phases of duration T_g, $g = 0, ..., G - 1$, and each sub-phase is divided into Q time slots of duration T_S, such that $T_g = QT_S$, $\forall g$. We assume that each time slot is indexed with index $gQ + t$, where $t = 0, ..., Q - 1$ and $g = 0, ..., G - 1$. The communication is organized on the sub-phase basis, as follows:

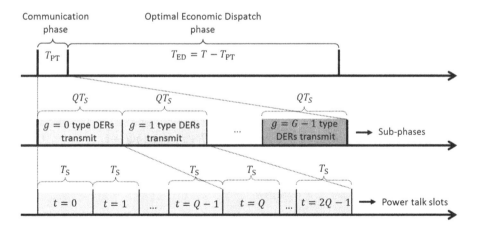

Fig. 2. Organization of the power talk communication protocol: phases, sub-phases and power talk slot.

---

[6] We note that the power talk communication in practice should also involve a channel estimation phase, where the DERs estimate the coefficients $\check{h}_{n,m}$; this aspect is out of the paper scope and we assume that the channel coefficients are known (see [17] for a detailed discussion on the channel estimation in power talk).

– All DERs of type g simultaneously transmit in sub-phase g, i.e., there are $|\mathcal{U}_g|$ transmitters in sub-phase g;
– All DERs of types $g, g+1, ..., G-1$ receive in sub-phase g; i.e., there are $\sum_{j=g}^{G-1} |\mathcal{U}_j|$ receivers in sub-phase g.

Obviously, DERs of type g work in full duplex mode in sub-phase g. By the end of sub-phase g, the DERs of type g will collect all aggregate generation capacities $w^{(0)}, ..., w^{(g)}$ that are required to run the decentralized OED (19).

When transmitting, DER u transmits a bit sequence (i.e., a word) that represents quantized value of its local generation capacity w_u, denoted by $\breve{w}_u$. On the other hand, the receiving units do not resolve the individual $\breve{w}_u$ for each $u \in \mathcal{U}_g$ that transmits simultaneously; rather, they detect the sum $w^{(g)}$ directly from the observations, as described below.

4.2   Detecting Aggregate Generation Capacities

Each transmitting DER $u \in \mathcal{U}_g$ quantizes the local value w_u with step $\triangle$ and 2^Q quantization levels, using the following rule:

$$\text{if } w_u \in [\beta_u \triangle, (\beta_u + 1)\triangle) \Rightarrow \breve{w}_u = \left(\beta_u + \frac{1}{2}\right)\triangle. \tag{20}$$

The quantization index $\beta_u \in \{0, ..., 2^Q - 1\}$ of the transmitting DER u is represented with uncoded stream of Q bits:

$$\{b_u(gQ + t) \in \{0, 1\}\}_{t=0,...,Q-1}, \ u \in \mathcal{U}_g. \tag{21}$$

Each bit $b_u(gQ + t)$ is then mapped into a corresponding power talk input (i.e., deviation of the reference voltage):

$$\Delta x_u(gQ + t) = \begin{cases} \lambda, & b_u(gQ + t) = 1, \\ -\lambda, & b_u(gQ + t) = 0. \end{cases} \tag{22}$$

A receiving DER k in slot $gQ + t$ of sub-phase g obtains the following measurement:

$$\Delta \tilde{v}'_k(gQ + t) = \Delta \tilde{v}_k(gQ + t) - \breve{h}_{n,n}\xi_{g,k}\Delta x_k(gQ + t) \tag{23}$$

$$= \sum_{m \in \mathcal{N}} \breve{h}_{n,m} \sum_{l \in \mathcal{U}_m, l \neq k} \xi_{g,l}\Delta x_l(gQ + t) + z_k(gQ + t) \tag{24}$$

$$= \lambda \sum_{m \in \mathcal{N}} \breve{h}_{n,m} \sum_{l \in \mathcal{U}_m, l \neq k} \xi_{g,l}\big(2b_l(gQ + t) - 1\big) + z_k(gQ + t). \tag{25}$$

Combining the measurements $\Delta \tilde{v}'_k(gQ+t)$, $t = 0, ..., Q-1$, collected in sub-phase g, the goal of the receiving DER k is to detect:

$$\breve{w}^{(g)} = \sum_{u \in \mathcal{U}_g, u \neq k} \breve{w}_u = \left(\sum_{u \in \mathcal{U}_g, u \neq k} \beta_u + \frac{|\mathcal{U}_g| - \xi_{g,k}}{2}\right)\triangle \tag{26}$$

$$= \left(\sum_{t=0}^{Q-1} \sum_{u \in \mathcal{U}_g, u \neq k} b_u(gQ + t)2^t + \frac{|\mathcal{U}_g| - \xi_{g,k}}{2}\right)\triangle. \tag{27}$$

Thus, the receiving DER k needs to determine the integer sum $\sum_{u \in \mathcal{U}_g} b_u(gQ + t) = \theta^{(g)}(gQ + t)$ of the bits received in slot $gQ + t$. The optimal Maximum-A-Posteriori (MAP) detection of $\theta^{(g)}(gQ + t)$ in slot $gQ + t$, is defined as follows:

$$\hat{\theta}^{(g)}(gQ + t) = \max_{\theta} f(\Delta\tilde{v}'_k(gQ + t); \theta^{(g)}(gQ + t) = \theta)\Pr(\theta^{(g)}(gQ + t) = \theta), \quad (28)$$

where $f(\Delta\tilde{v}'_k(gQ+t); \theta^{(g)}(gQ+t) = \theta)$ is the density function of the measurement $\Delta\tilde{v}'_k(gQ + t)$, parametrized w.r.t. the integer sum $\theta = 0, ..., |\mathcal{U}_g| - \xi_{g,k}$, and $\Pr(\theta^{(g)}(gQ+t) = \theta)$ is the *a priori* probability of θ. Note that the maximization is performed w.r.t. $\theta = 0, ..., |\mathcal{U}_g| - \xi_{g,k}$, which implies that the complexity of the detector grows linearly with the number of simultaneously transmitting DERs in a sub-phase. Under Gaussian noise assumption, (28) becomes:

$$\hat{\theta}^{(g)} = \max_{\theta} \sum_{j=1}^{\binom{|\mathcal{U}_g| - \xi_{g,k}}{\theta}} \exp\left\{ -\frac{(\Delta\tilde{v}'_k - \lambda\sum_{m \in \mathcal{N}} \check{h}_{n,m} \sum_{l \in \mathcal{U}_m, l \neq k} \xi_{g,l}(2b_l^j - 1))^2}{2\sigma^2} \right\},$$
$$(29)$$

where the slot index $gQ+t$ is omitted due to space limitation. The summation in (29) is over all $\binom{|\mathcal{U}_g| - \xi_{g,k}}{\theta}$ binary sequences $b_l^j(gQ + t)$, $j = 0, ..., \binom{|\mathcal{U}_g| - \xi_{g,k}}{\theta} - 1$ of length $|\mathcal{U}_g| - \xi_{g,k}$. These sums are computed by each receiving DER and stored in memory for each $\theta = 0, ..., |\mathcal{U}_g| - \xi_{g,k}$ and each g. Using (29), a DER receiving in sub-phase g detects the aggregate generation capacity:

$$\hat{w}^{(g)} = \left(\sum_{t=0}^{Q-1} \hat{\theta}^{(g)}(gQ + t)2^t + \frac{|\mathcal{U}_g| - \xi_{g,k}}{2} \right) \triangle . \quad (30)$$

In the case when $G = U$, i.e., the costs per unit output power are different for all DERs, the above power talk strategy reduces to simple TDMA solution, where a single DER transmits in each sub-phase.

Finally, we provide a policy for choosing the parameter λ. We constrain the variance of the output power deviations of each DER as follows:

$$\text{Var}(\Delta p_u) \leq \pi^2, \ u \in \mathcal{U}, \quad (31)$$

where π is the power deviation budget of each unit, corresponding to the system tolerance to power deviations. This constraint yields the following range for λ:

$$0 < \lambda \leq \min_{u} \left\{ \frac{\pi}{\sqrt{\sum_{m \in \mathcal{N}} \check{\phi}_{u,m}^2 \sum_{l \in \mathcal{U}_m} \xi_{g,l}}} \right\}, \quad (32)$$

where we used the linear approximation (14) for Δp_u.

4.3   Cost Trade-Off

At the end of the communication phase, each DER $u \in \mathcal{U}$ operates with imperfect knowledge of the sum generation capacities $w^{(g)}$, due to quantization and detection error. The corresponding, potentially suboptimal dispatch policy vector obtained via (19) is denoted with $\hat{\mathbf{p}}^\star$. The total generation cost when the policy $\hat{\mathbf{p}}^\star$ is enforced, is denoted with $\hat{C}^\star$. Further, a dispatch policy might under- or over-estimate the cumulative generation capacities, leading to power deficit $p^{\mathrm{def}} = (d^{\mathrm{L}} - \sum_{g=0}^{G-1} \sum_{u \in \mathcal{U}_g} \hat{p}_u^\star)^+$ or power surplus $p^{\mathrm{sur}} = (\sum_{g=0}^{G-1} \sum_{u \in \mathcal{U}_g} \hat{p}_u^\star - d^{\mathrm{L}})^+$. In the case of deficit, a back-up source (e.g., a storage system or supply from the main grid) is activated; the cost per unit generation of the back-up source is denoted with c^{def}. Similarly, the power surplus is transferred to storage system/main grid at cost $c^{\mathrm{sur}} > 0$ [6,8].

The total generation cost for a dispatch period can be calculated as follows:

$$C(\hat{\mathbf{p}}^\star) = \left(1 - \frac{T_{\mathrm{S}}}{T} QG\right) \underbrace{(\hat{C}^\star + \varrho)}_{\hat{\Omega}^\star} + \frac{T_{\mathrm{S}}}{T} \sum_{g=0}^{G-1} \sum_{t=0}^{Q-1} \sum_{g=0}^{G-1} \sum_{u \in \mathcal{U}_g} c_g p_u(gQ + t) \tag{33}$$

$$= \hat{\Omega}^\star + \frac{T_{\mathrm{S}}}{T} QG \left(\sum_{g=0}^{G-1} \sum_{u \in \mathcal{U}_g} c_g (\overline{p}_u - \hat{p}_u^\star) - \varrho \right), \tag{34}$$

where $\varrho = c^{\mathrm{def}} p^{\mathrm{def}} + c^{\mathrm{sur}} p^{\mathrm{sur}}$, and where we assumed that $p_u(gQ + t) = \overline{p} + \Delta p_u(gQ + t)$ and $\lim_{Q \to \infty} \sum_{t=0}^{Q-1} \Delta p_u(gQ + t) = 0$, i.e., it is assumed that antipodal signaling, see (22), results in (roughly) symmetric supply power deviations around $\overline{p}_u$. Equation (34) provides insight into the fundamental trade-off of the proposed solution. Namely, $\lim_{Q \to \infty, T_{\mathrm{S}} \to \infty} \hat{\Omega}^\star = C^\star < \hat{\Omega}^\star$; however, increasing Q and/or the slot duration, increases the duration of the communication phase where the system operates suboptimally, potentially increasing the overall cost.

5 Evaluation

In this section, we evaluate the performance of the proposed technique by simulating a single bus system (i.e., $N = 1$), to which all DERs and loads are connected through lines with negligible resistances. This is a valid model for small, localized MGs, where the effect of the transmission network on the power flow is negligible [6,8]. There are $U = 10$ DERs, organized into $G = 4$ types:

$$\Xi = \begin{bmatrix} 1\,1\,1\,0\,0\,0\,0\,0\,0\,0 \\ 0\,0\,0\,1\,1\,0\,0\,0\,0\,0 \\ 0\,0\,0\,0\,0\,1\,1\,1\,0\,0 \\ 0\,0\,0\,0\,0\,0\,0\,0\,1\,1 \end{bmatrix}. \tag{35}$$

The cost vector is $\mathbf{c} = [5, 5, 5, 7.5, 7.5, 10, 10, 10, 50, 50]$ and $c^{\mathrm{def}} = c^{\mathrm{sur}} = 100$; note that these numbers are used only for illustrative purposes. The power generation of each DG changes uniformly and independently in the interval $w_u \in [0, w_{\mathrm{max}} = 2\,\mathrm{kW}]$, while the total load power demand is $d^{\mathrm{L}} = 5\,\mathrm{kW}$; the load is composed only of constant power part, i.e., $d^{\mathrm{L}} = d^{\mathrm{cp}}$. The quantization

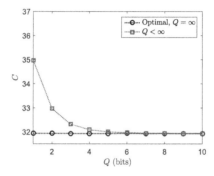

Fig. 3. Optimal economic dispatch: effect of quantization error.

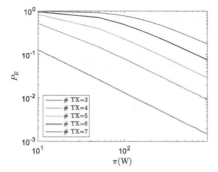

Fig. 4. Error probability of the detector (29), parametrized by the number of simultaneously transmitting DERs.

step is $\triangle = \frac{w_{max}}{2^Q}$. The total duration of the dispatch period is fixed to $T = 300\,\mathrm{s}$. The sampling frequency of the converter's front-end is $f_S = 50$ kHz, and the standard deviation of the voltage sampling noise is $\sqrt{N_0} = 0.1$ V/sample [17]. We investigate the cost behavior as a function of the number of bits Q for varying slot durations T_S and different tolerances π on the output power deviations.

First, we illustrate the effect of the quantization error on the optimal decentralized dispatch policy (19), presented in Fig. 3. It can be noted that this effect becomes negligible for $Q > 10$. This implies that in usual MG control applications, the length of the messages that need to be exchanged among the units is exceptionally short. In fact, in virtually all upper level control applications, around 2 bytes $(= 16$ bits$)$ of information per node message is sufficient [6].

Next, we investigate the performance of the detector (29). The detector operates with averages over single power talk slot, i.e., the observed value $\Delta\tilde{v}_u(gQ+t)$ is obtained by averaging $T_S f_S$ samples during the slot $gQ+t$. Therefore, the standard deviation of the noise component $z_u(gQ+t) \sim \mathcal{N}(0,\sigma^2)$ in each slot, is $\sigma = \sqrt{\frac{N_0}{T_S f_S}}$. Figure 4 shows the average probability of error P_E of the detector (29) using $T_S = 0.1$ s and $Q = 10$ as a function of π, for increasing number of simultaneously transmitting units. Obviously, the probability of error increases as the number of units increases. On the other hand, applications that can tolerate larger output voltage deviations, i.e., larger π, can benefit from improved detector performance. It can be concluded that our detector is well suited for MG control applications as in typical MG setup the total number of DERs is low, typically less than ~ 10.

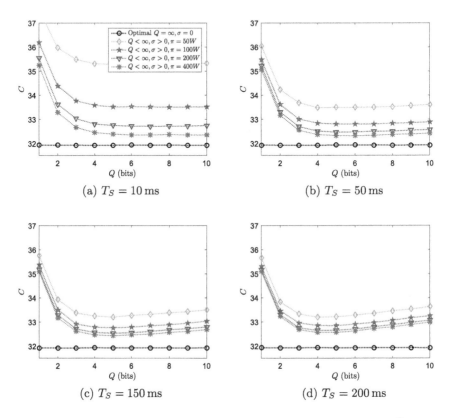

(a) $T_S = 10$ ms

(b) $T_S = 50$ ms

(c) $T_S = 150$ ms

(d) $T_S = 200$ ms

Fig. 5. Generation cost of the optimal decentralized dispatch policy $\hat{\tilde{p}}^\star$.

Figure 5 illustrates the total generation cost of the dispatch policy $\hat{\tilde{p}}^\star$ (34). Increasing T_S suppresses the noise (recall that $\sigma = \sqrt{\frac{N_0}{T_S f_S}}$), which improves the detector performance and pushes the first term in (34) towards the optimal value. At the same time, the overall duration of the power talk phase is increased, which increases the value of the second term. One way to improve the first term in (34) while keeping the second term fixed is through increasing π; however, the choice of π will be determined by the specific amount of deviation of the electric parameters that can be tolerated in the system. For instance, in LVDC MG system with rated voltage of 400 V, reference voltage deviations of ± 2 V amount to $\pm 0.5\%$ deviation from the rated value; these deviations preserve the small signal assumption (10) and allow values for π of up to 200 W, which significantly improves the performance of the detector, see Fig. 4. We conclude that power talk indeed shows strong potential as a communication enabler for upper layer control applications in MGs.

6   Concluding Remarks

We presented a powerline communication protocol for control applications in DC MicroGrids, specifically designed to facilitate distributed optimal economic dispatch without support of an external communication network. In the proposed solution, the distributed generators, transmit information about their local, instantaneous generation capacities over the power lines in full duplex mode, while the receiving generators use specific integer sum detector to retrieve the aggregate generation capacity of the transmitting generators. On the physical layer, the solution exploits the multiple access nature of the power talk communication channel in which, information is modulated into the parameters of the primary control loops of power electronic converters. The simulation results illustrate the inherent trade-offs and prove that the propose solution is a viable communication alternative for self-sustainable and self-sufficient DC MG systems.

References

1. Zubieta, L.E.: Are microgrids the future of energy?: Dc microgrids from concept to demonstration to deployment. IEEE Electrification Mag. **4**(2), 37–44 (2016)
2. Dragicevic, T., Lu, X., Vasquez, J.C., Guerrero, J.M.: DC microgrids; part I: a review of control strategies and stabilization techniques. IEEE Trans. Power Electron. **31**(7), 4876–4891 (2016)
3. Jin, C., Wang, P., Xiao, J., Tang, Y., Choo, F.H.: Implementation of hierarchical control in DC microgrids. IEEE Trans. Industr. Electron. **61**(8), 4032–4042 (2014)
4. Zhao, J., Dorfler, F.: Distributed control and optimization in DC microgrids. Automatica **61**, 18–26 (2015)
5. Dragicevic, T., Guerrero, J.M., Vasquez, J.C., Skrlec, D.: Supervisory control of an adaptive-droop regulated DC microgrid with battery management capability. IEEE Trans. Power Electron. **29**(2), 695–706 (2014)
6. Liang, H., Choi, B.J., Abdrabou, A., Zhuang, W., Shen, X.S.: Decentralized economic dispatch in microgrids via heterogeneous wireless networks. IEEE J. Sel. Areas Commun. **30**(6), 1061–1074 (2012)
7. Gan, L., Low, S.H.: Optimal power flow in direct current networks. IEEE Trans. Power Syst. **29**(6), 2892–2904 (2014)
8. Giannakis, G.B., Kekatos, V., Gatsis, N., Kim, S.J., Zhu, H., Wollenberg, B.F.: Monitoring and optimization for power grids: a signal processing perspective. IEEE Sig. Process. Mag. **30**(5), 107–128 (2013)
9. Schonberger, J., Duke, R., Round, S.D.: DC-bus signaling: a distributed control strategy for a hybrid renewable nanogrid. IEEE Trans. Industr. Electron. **53**(5), 1453–1460 (2006)

10. Chen, D., Xu, L., Yao, L.: DC voltage variation based autonomous control of DC microgrids. IEEE Trans. Power Deliv. **28**(2), 637–648 (2013)
11. Vandoorn, T.L., Renders, B., Degroote, L., Meersman, B., Vandevelde, L.: Active load control in islanded microgrids based on the grid voltage. IEEE Trans. Smart Grid **2**(1), 139–151 (2011)
12. Angjelichinoski, M., Stefanovic, C., Popovski, P., Liu, H., Loh, P.C., Blaabjerg, F.: Power talk: how to modulate data over a DC micro grid bus using power electronics. In: IEEE Global Communications Conference (2015)
13. Angjelichinoski, M., Stefanovic, C., Popovski, P., Blaabjerg, F.: Power talk in DC micro grids: constellation design and error probability performance. In: IEEE International Conference on Smart Grid Communications, pp. 689–694 (2015)
14. Angjelichinoski, M., Stefanovic, C., Popovski, P.: Power talk for multibus DC microGrids: creating and optimizing communication channels. In: IEEE Global Communications Conference (2016, to appear)
15. Galli, S., Scaglione, A., Wang, Z.: For the grid and through the grid: the role of power line communications in the smart grid. Proc. IEEE **99**(6), 998–1027 (2011)
16. Angjelichinoski, M., Scaglione, A., Popovski, P., Stefanovic, C.: Distributed estimation of the operating state of a single-bus DC microgrid without an external communication network. In: IEEE Global Conference on Signal and Information Processing (2016, to appear)
17. Angjelichinoski, M., Stefanovic, C., Popovski, P., Liu, H., Loh, P.C., Blaabjerg, F.: Multiuser communication through power talk in DC microgrids. IEEE J. Sel. Areas Commun. **34**(7), 2006–2021 (2016)
18. Angjelichinoski, M., Stefanovic, C., Popovski, P., Blaabjerg, F.: Communication-theoretic model of power talk for a single-bus DC microgrid. Information **7**(1), 18 (2016)

On the Impact of Precoding Errors on Ultra-Reliable Communications

Guillermo Pocovi[1(✉)], Klaus I. Pedersen[1,2], and Beatriz Soret[2]

[1] Aalborg University, Aalborg, Denmark
gapge@es.aau.dk
[2] Nokia Bell Labs, Aalborg, Denmark

Abstract. Motivated by the stringent reliability required by some of the future cellular use cases, we study the impact of precoding errors on the SINR outage performance for various spatial diversity techniques. The performance evaluation is carried out via system-level simulations, including the effects of multi-user and multi-cell interference, and following the 3GPP-defined simulation assumptions for a traditional macro case. It is shown that, except for feedback error probabilities larger than 1%, closed-loop microscopic diversity schemes are generally preferred over open-loop techniques as a way to achieve the SINR outage performance required for ultra-reliable communications. Macroscopic diversity, where multiple cells jointly serve the UE, provides additional robustness against precoding errors. For example, a 4×4 MIMO scheme with two orders of macroscopic diversity can achieve the 0 dB SINR outage target at the 10^{-5}-th percentile, even for a precoding error probability of 1%. Based on the obtained results, it is discussed what transmission modes are more relevant depending on the feedback error constraint.

1 Introduction

Ultra-reliable communications over wireless is an active research topic that will open the possibility of novel applications [1]. For some of the use cases, latencies of a few milliseconds must be guaranteed with reliability levels up to 99.999%. The signal to interference-and-noise ratio (SINR) outage performance is a relevant metric for ultra-reliable communications. In this context, spatial diversity techniques such as microscopic and macroscopic diversity have shown promising potential. For example, the work in [2,3] shows that the proper combination of macroscopic and microscopic diversity techniques can provide the required SINR outage performance.

Microscopic diversity is typically used in modern cellular systems, such as the Long Term Evolution (LTE), by use of multiple-input multiple-output (MIMO) antenna techniques. In the downlink, the gains provided by microscopic diversity strongly depend on the availability and accuracy of channel state information (CSI) at the eNodeB. If the channel knowledge is precise enough, closed-loop (CL) schemes, which are known to provide the best performance [4], can be applied.

© Springer International Publishing AG 2016
T.K. Madsen et al. (Eds.): MACOM 2016, LNCS 10121, pp. 45–54, 2016.
DOI: 10.1007/978-3-319-51376-8_4

However, in cases of absence or inaccurate CSI knowledge due to e.g. imperfect channel estimation, open-loop (OL) schemes are typically more appropriate.

In frequency division duplex (FDD) modes, where channel reciprocity is not applicable, the eNodeB obtains the CSI through an uplink feedback channel. The CSI contains information about the current channel quality, and the preferred precoding matrix to be applied in downlink CL transmissions. Apart from the typically applied quantization in order to cope with the limited feedback capacity of real systems, the precoding information is prone to errors due to the intrinsic presence of fading and interference in the wireless channel. The impact of CSI feedback errors have been evaluated from a system capacity point of view. For example, the work in [5] evaluates the influence of CSI feedback errors on the throughput performance of multi-user MIMO systems, whereas [6] demonstrates the significant performance degradation when a UE intentionally reports the wrong CSI to the eNodeB. Previous reliability analyses [2,3] have not considered these types of imperfections. Our hypothesis is that precoding errors could have a significant impact on ultra-reliable communications, which is what we evaluate in this work.

In this paper we study the impact of CSI feedback errors on the achievable downlink SINR performance in a multi-cell multi-user environment. Our focus is on the very-low percentiles of the SINR distribution in order to quantify the impact of feedback errors on ultra-reliable communications, and determine what transmission modes (e.g. OL or CL) are more relevant depending on the feedback error probability. The complexity of our system model prevents a purely analytical evaluation without omitting important aspects influencing the performance. The evaluation is carried out following the 3GPP-defined simulation assumptions for a LTE macro cellular network that relies on commonly accepted models and methodologies. Mathematical expressions for the user-experienced SINR, when applying the different transmission schemes and related imperfections, are presented in this article and used in the simulations. Long simulations are run to ensure statistical reliable performance results with high level of confidence

The rest of the paper is outlined as follows: Sect. 2 describes our system model. The simulation assumptions are outlined in Sect. 3. Performance results are presented in Sect. 4, followed by concluding remarks in Sect. 5.

2   System Model

The network consists of a set of $\mathcal{N} = \{1, ..., N\}$ cells, each equipped with T transmit antennas, and a set of $\mathcal{K} = \{1, ..., K\}$ UEs with R receive antennas. Each downlink connection between UE $k \in \mathcal{K}$ and its serving cell $j \in \mathcal{N}$ is represented by a TxR CL MIMO system as shown in Fig. 1. As our focus is on reliable communications, only single-stream transmission cases are considered [7]. First, each UE estimates the RxT-dimensional channel $\mathbf{H}_{jk}$, whose (m,n)-th element represents the complex channel gain from transmit antenna n at cell j, to receive antenna m at UE k. As a second step, the vector $\mathbf{u}_j$ corresponding to the largest eigenvalue of the $\mathbf{H}_{jk}^H \mathbf{H}_{jk}$ matrix is calculated through singular

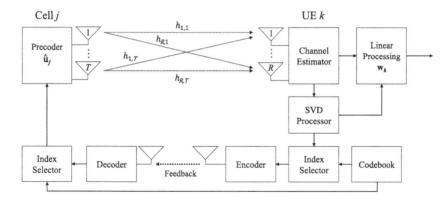

Fig. 1. Transmitter-receiver architecture.

value decomposition (SVD), i.e. $\mathbf{u}_j = \mathrm{EIG}_{max}(\mathbf{H}_{jk}^H \mathbf{H}_{jk})$. Next, an index selector selects from a pre-defined codebook the precoding vector that matches best with $\mathbf{u}_j$. We refer to this quantized version as $\hat{\mathbf{u}}_j$. The index to the precoder, i.e. precoding matrix indicator (PMI), is transmitted to the cell through the uplink feedback channel.

The cell uses the received PMI to obtain $\hat{\mathbf{u}}_j$, which is then applied in the data transmission. Within each cell, the UEs are served on orthogonal resources, i.e. there is no intra-cell interference as it is also the case for LTE assuming single-stream and single-user MIMO transmission modes [4]. In a frequency-flat fading case, the R-dimensional received signal $\mathbf{r}_j$ by a user (for simplicity, we omit the user-specific index) served in cell $j \in \mathcal{N}$ is given as follows,

$$\mathbf{r}_j = \mathbf{H}_j \sqrt{\Omega_j} \hat{\mathbf{u}}_j s_j + \sum_{i \in \mathcal{N} \backslash j} \mathbf{H}_i \sqrt{\Omega_i} \hat{\mathbf{u}}_i s_i + \mathbf{n} \, , \tag{1}$$

where Ω_i represents the averaged received power from the i-th cell, including the effect of the antenna gain and pattern, distance-dependent attenuation and shadowing; s_i represents the transmitted symbol (for simplicity, $\|s_i\| = 1$) and $\mathbf{n}$ is a Rx1 zero mean Gaussian vector with variance σ^2 representing the noise power at each receiving antenna.

In order to maximize the received signal power at the receiver, the R received signals are combined by applying a weight vector $\mathbf{w} = \mathbf{H}_j \hat{\mathbf{u}}_j$. The resulting post-detection SINR expression is given by,

$$SINR_j = \frac{\Omega_j \|\hat{\mathbf{u}}_j^H \mathbf{H}_j^H \mathbf{H}_j \hat{\mathbf{u}}_j\|^2}{\sum_{i \in \mathcal{N} \backslash j} \Omega_i \|\hat{\mathbf{u}}_j^H \mathbf{H}_j^H \mathbf{H}_i \mathbf{u}_i\|^2 + \sigma^2 \|\hat{\mathbf{u}}_j^H \mathbf{H}_j^H\|^2} \, , \tag{2}$$

where $[\cdot]^H$ denotes the Hermitian transpose.

The presented microscopic scheme corresponds to transmission mode 6 (TM6) in LTE terminology. TM6 is a special case of CL spatial multiplexing (TM4) where the transmission rank is limited to one. The UE utilizes the down-link cell-specific reference signals (RS) to perform the channel estimation, and select the preferred PMI (from a common codebook). The eNodeB signals the applied precoding to the UE in the downlink grant [4].

TM6 allows operation with 2 or 4 transmit antennas. For the former case, the LTE Release 8 codebook contains 4 precoding vectors, whereas there are 16 different entries for four transmit antennas [8]. The number of entries have been selected as a tradeoff between the uplink signalling overhead and downlink performance.

In cases where channel information is missing at the eNodeB, spatial diversity gain can be obtained with open-loop transmission modes. LTE transmission mode 3 (TM3) supports OL spatial diversity by use of space-frequency block coding (SFBC) techniques [9], which are based on the space-time block coding initially proposed by Alamouti [10]. SFBC achieves similar diversity order to CL, but with a reduced received power since the transmit beamforming gain is not obtained [10]. The post-detection SINR of a TxR OL MIMO scheme is simply modelled by adding a $10\log_{10} T$ SINR penalty to the performance obtained with a TxR MIMO system assuming full channel knowledge at the transmitter (i.e. without quantized precoding) [10].

As a method to further improve the SINR outage performance, we also consider macroscopic diversity transmissions from M cells to a certain UE [2]. We assume a simple soft-combining approach as known from Universal Mobile Telecommunications System (UMTS), where the received signal from each macroscopic branch is independently detected and combined at the UE [11]. As this scheme rely on non-coherent transmissions, each of the M macroscopic links can be modelled as shown in Fig. 1. The SINR after combining M ($1 \leq M \leq N$) macroscopic branches is expressed as follows,

$$SINR = \sum_{j=1}^{M} SINR_j \,, \tag{3}$$

where $SINR_j$ is the SINR calculated according to (2), assuming the UE is connected to cell j.

2.1   Precoding Errors

The gains provided by spatial diversity techniques depends on the accuracy of the CSI at the transmitter [4]. Since the CSI is estimated at the UE and transmitted to the eNodeB through an uplink feedback channel, it is vulnerable to multiple sources of delay and other imperfections. The delays are a consequence of the constrained CSI reporting periodicity and processing time, meaning that the optimal precoding will not be immediately applied at the transmitter. Additionally, errors in the channel estimation could lead to a sub-optimal PMI selection. Another source of degradation is errors in the uplink transmission of the CSI due to the inevitable presence of fading and interference in the wireless channel.

We focus uniquely on the effect of precoding feedback errors. We assume that errors in the feedback channel can occur with a given error probability P_e. In such cases, the PMI decoded by the eNodeB will be different to the reported by the UE, which will lead to a erroneous precoder selection. The errors in the feedback channel are assumed to be i.i.d for each UE-eNodeB connection.

Since the eNodeB signals the applied precoding in the scheduling grant, the UE can still apply a proper combining weight vector to improve the signal quality at the receiver. In other words, the benefits of transmit diversity are lost but the receive diversity gain is maintained. As a more pessimistic case, errors could alter the applied-PMI related signalling in the downlink grant, resulting in loss of both the transmit and receive diversity gain.

3   Simulation Assumptions

The evaluation is carried out by analysing the downlink SINR distribution for different antenna schemes, transmission methods, and feedback error probabilities. A snapshot-based simulation approach is applied and the respective assumptions are summarized in Table 1. A large macro-cellular network composed of three-sector sites with inter-site distance of 500 m is assumed, where UEs are uniformly distributed [12]. Cells are transmitting at full power (full load conditions) at a 2 GHz carrier frequency. The simulation procedure is as follows: Each UE selects M serving cells according to the average received power. Effects of user mobility and handovers are not explicitly included in the simulations. However, the effect of handover hysteresis margin is implicitly modelled in the active set selection algorithm: each UE identifies the strongest received cells that are within a certain *handover window*, as compared to the strongest cell. A serving cell for the UE is then randomly selected from the cells within the handover window. This method models the effect where not all UEs are served by their strongest cell due to the use of handover hysteresis margins in reality.

The experienced instantaneous post-detection SINR is calculated for each UE following the models in Sect. 2. For each snapshot, the fast fading is independent and identically distributed for each transmit-receive antenna pair, following a complex Gaussian distribution (i.e. the envelope is Rayleigh distributed). Additive white Gaussian noise with a power spectral density of -174 dBm/Hz is considered. It is assumed that UEs are scheduled with 10 MHz bandwidth, resulting in a noise power of -96 dBm when including a 8 dB noise figure at the UE.

A large number of snapshots are simulated and the generated SINR samples are used to form empirical cumulative distribution functions (CDF). Our target is to study the impact of different feedback error probabilities on the SINR outage performance. In line with [1], the key performance indicator (KPI) is the SINR at the 10^{-5}-th percentile. At this percentile, we consider a 0 dB SINR as an appropriate target to have error-free downlink reception, and therefore fulfil the low latency requirements of ultra-reliability use cases (we refer to [2] for more details).

Table 1. Simulation assumptions

Parameter	Value
Network layout	3GPP Macro case 1
UE distribution	Uniformly distributed in outdoor locations
Macro cell transmit power	46 dBm
Carrier frequency	2.0 GHz
Propagation	$128.1 + 37.6 \log 10(R[\text{km}])$ dB
Antenna gain	BS: 14 dBi. UE: 0 dBi
Antenna pattern	BS: 3D with 12° downtilt UE: omnidirectional
Shadowing distribution	Log-normal with $\sigma = 8$ db
Shadowing correlation	Intra-site: 1.0; Inter-site: 0.0
Noise power spectral density	-174 dBm/Hz
Noise figure	8 dB
Noise power	-96 dBm @10 MHz
Handover window	3 dB
Fast fading	Rayleigh distributed; Uncorrelated among the different antenna branches
Feedback error probability P_e	10^{-1}, 10^{-2}, 10^{-3}
SINR outage target	0 dB at the 10^{-5}-th percentile

4 Results

The first set of results correspond to the relatively pessimistic case where the PMI applied by the eNodeB is unknown by the UE, thus the UE assumes that the applied precoding is the one that it has previously signalled. Figure 2 shows the empirical CDF of the SINR distribution for 2×2 and 4×4 schemes, OL and CL transmission modes, and different feedback error probabilities. Obviously, the 4×4 schemes offer superior performance as compared to 2×2 MIMO schemes. The benefits of CL transmissions over OL schemes are also observable: 4.6 dB and 2.2 dB SINR gain for 2×2 and 4×4 schemes, respectively, at the 10^{-5}-th percentile.

When including the effects of feedback errors, a significant degradation of the performance is observed. For example, even for $P_e = 10^{-3}$, the experienced SINR degradation at the 10^{-5}-th percentile is as high as 8.9 dB and 3.2 dB for 2×2 and 4×4 antenna schemes. The reason is that, when this type of errors occur, the benefits of both transmit and receive diversity are not obtained, i.e. the instantaneously experienced diversity order is equivalent to a 1×1 MIMO system. Under such circumstances, it is shown how OL schemes, which do not require any uplink CSI feedback, offer better performance.

Next, we consider the case where the eNodeB applies an erroneous precoding vector, but the applied PMI is known at the receiver. Figures 3 and 4 shows the SINR distribution with 2×2 and 4×4 antenna schemes, respectively. Cases

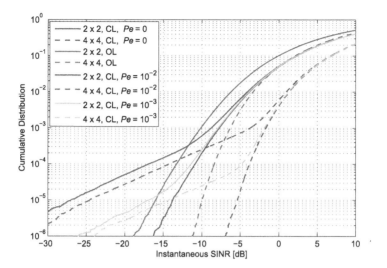

Fig. 2. SINR outage performance with 2×2 and 4×4 antenna schemes, different transmission modes and precoding error probabilities (P_e). $M = 1$.

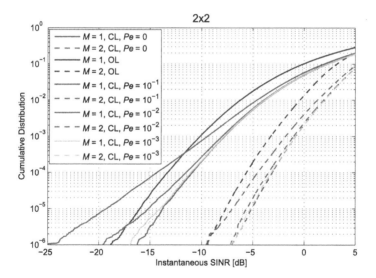

Fig. 3. SINR outage performance with a 2×2 antenna scheme, different transmission modes and precoding error probabilities (P_e). It is assumed that the applied PMI is known at the UE.

with second order of macroscopic diversity and $P_e = 10^{-1}$ are also shown. As compared to the performance results in Fig. 2, errors in the uplink feedback have less impact on the SINR performance. For example, CL configurations with $P_e = 10^{-3}$ and $M = 1$ experience a performance degradation of only 0.3 dB. In this case, the receiver has knowledge of the applied precoding, which allows to

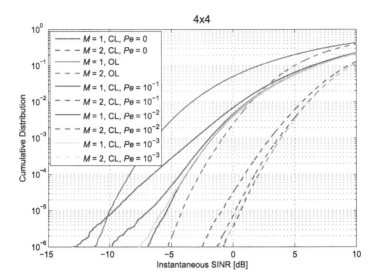

Fig. 4. SINR outage performance with a 4×4 antenna scheme, different transmission modes and precoding error probabilities (P_e). It is assumed that the applied PMI is known at the UE.

fully harvest the receive diversity gain. It is observed that for $P_e \leq 10^{-2}$, the performance of CL schemes is better than OL.

As also observed in previous studies [2], macroscopic diversity provides additional protection against fast and slow fading hence providing significantly better SINR performance. Even for $P_e = 10^{-1}$, only a 1.3 dB performance degradation is observed for 2×2 and 4×4 MIMO schemes. With macroscopic diversity, the probability of experiencing feedback error across the M links is reduced. Note that compared to the intra-cell MIMO schemes, the considered macroscopic diversity technique relies on non-coherent transmissions and soft-combining of the multiple received signals at the UE, therefore it is only required to report traditional CSI feedback to each of the M eNodeBs.

Figure 5 summarizes the achieved 10^{-5}-th percentile SINR performance under different transmission schemes and feedback error probabilities. The 0 dB SINR target is represented with a horizontal dashed line. As also concluded in [2], a 4×4 CL MIMO scheme with $M = 2$ allows to fulfil the 0 dB SINR target. However, it is observed that this is only achievable under certain feedback error probabilities. For instance, if the feedback error probability is $P_e \geq 10^{-1}$, 4×4 MIMO with $M = 2$ no longer fulfils the 0 dB SINR target. The SINR degradation due to feedback errors is much more severe for configurations with low diversity order. For example, 4×4 CL MIMO with $M = 1$ achieves similar performance as 4×4 OL for $P_e = 10^{-1}$; whereas, under the same error probability, 2×2 CL with $M = 1$ is 3.2 dB worse than OL.

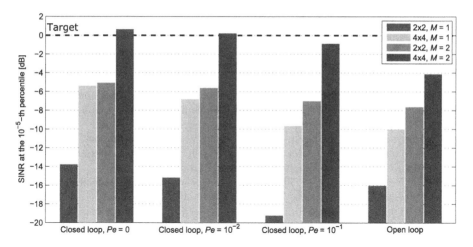

Fig. 5. Achieved SINR at the 10^{-5}-th percentile for several transmission schemes and precoding error probabilities (P_e).

5   Conclusions

In this paper we have evaluated the SINR outage performance under different CSI feedback error constraints in order to quantify its impact on ultra-reliable communications. It have been shown that even for feedback error probabilities as high as 10^{-2} (i.e. three orders of magnitude larger than the required reliability), there is a benefit of using closed-loop MIMO schemes over open-loop schemes. The performance degradation due to errors in the feedback can be reduced by applying macroscopic diversity, as the considered scheme relies on non-coherent independent transmissions from the different macroscopic branches. For instance, a 4×4 MIMO scheme with two orders of macroscopic diversity can achieve the 0 dB SINR outage target at the 10^{-5}-th percentile, even with a 1% error probability in the CSI feedback. For configurations with low diversity order, a larger performance impact has been observed. For example, closed-loop 2×2 MIMO without macroscopic diversity, performs 3.2 dB worse than open-loop transmissions for a 10% feedback error probability. Future work must also consider other sources of imperfections in the channel information. For instance, as a consequence of non-ideal channel estimation at the UE, or due to delays in the CSI report. This will allow to fully assess the reliability performance in a practical setting.

References

1. 3GPP TR 38.913 v0.3.0. Study on scenarios and requirements for next generation access technologies, March 2016
2. Pocovi, G., Soret, B., Lauridsen, M., Pedersen, K.I., Mogensen, P.: Signal quality outage analysis for ultra-reliable communications in cellular networks. In: IEEE Globecom Workshops, December 2015

3. Kirsten, F., Ohmann, D., Simsek, M., Fettweis, G.P.: On the utility of macro- and microdiversity for achieving high availability in wireless networks. In: IEEE PIMRC, September 2015

4. Holma, H., Toskala, A.: LTE for UMTS: Evolution to LTE-Advanced. Wiley, Hoboken (2011)

5. Mielczarek, B., Krzymień, W.A.: Influence of CSI feedback errors on capacity of linear multi-user MIMO systems. In: IEEE Vehicular Technology Conference, April 2007

6. Mukherjee, A., Swindlehurst, A.L.: Poisoned feedback: The impact of malicious users in closed-loop multiuser MIMO systems. In: IEEE International Conference on Acoustics Speech and Signal Processing, March 2000

7. Tse, D.N.C., Viswanath, P., Zheng, L.: Diversity-multiplexing tradeoff in multiple-access channels. IEEE Trans. Inf. Theor. **50**(9), 1859–1874 (2004)

8. 3GPP TS 36.211 v12.5.0. Evolved Universal Terrestrial Radio Access (E-UTRA); Physical channels and modulation, April 2014

9. Dehghani, M.J., Aravind, R., Jam, S., Prabhu, K.M.: Space-frequency block coding in OFDM systems. In: IEEE TENCON, November 2004

10. Alamouti, S.: A simple transmit diversity technique for wireless communication. IEEE J. Select. Areas Commun. **16**(8), 1451–1458 (1998)

11. Holma, H., Toskala, A. (eds.): WCDMA for UMTS - Radio Access for Third Generation Mobile Communications, 3rd edn. Wiley, Hoboken (2004)

12. 3GPP TR 36.814 v9.0.0. Evolved Universal Terrestrial Radio Access (E-UTRA); Further advancements for E-UTRA physical layer aspects, March 2010

MAC Layer Aspects

Joint Usage of Dynamic Sensitivity Control and Time Division Multiple Access in Dense 802.11ax Networks

Evgeny Khorov[1,2(✉)], Anton Kiryanov[2], Alexander Krotov[2], Pierluigi Gallo[3], Domenico Garlisi[3], and Ilenia Tinnirello[3]

[1] Skoltech, Moscow, Russia
e@khorov.ru
[2] IITP RAS, Moscow, Russia
[3] DEIM - Università di Palermo/CNIT, Palermo, Italy
pierluigi.gallo@unipa.it

Abstract. It is well known that in case of high density deployments, Wi-Fi networks suffer from serious performance impairments due to hidden and exposed nodes. The problem is explicitly considered by the IEEE 802.11ax developers in order to improve spectrum efficiency. In this paper, we propose and evaluate the joint usage of dynamic sensitivity control (DSC) and time division multiple access (TDMA) for improving the spectrum allocation among overlapping 802.11ax BSSs. To validate the solution, apart from simulation, we used a testbed based on the Wireless MAC Processor (WMP), a prototype of a programmable wireless card.

Keywords: IEEE 802.11ax · Dynamic sensitivity control · TDMA · Dense deployment · Hidden node problem · Exposed node problem

1 Introduction

In recent years, it has clearly emerged that Wi-Fi performance can dramatically degrade in multi-hop and high-dense scenarios. These conditions are likely to occur when multiple networks coexist [1,2] or a large access infrastructure is deployed [3,4]. The main reasons for the degradation include the starvation and unfairness phenomena of CSMA-based protocols due to a mismatch in the local views of the wireless medium among the nodes, which is only partially solved by the RTS/CTS handshake [5].

The problem of performance degradation is becoming increasingly critical because of the popularity of Wi-Fi network deployments, especially in residential areas, business offices, and indoor/outdoor hotspots, e.g. mass events, airports, shopping malls. Because of this problem, the IEEE 802 LAN/MAN Standards Committee (LMSC) is working on a novel 802.11ax amendment [6], which explicitly aims to optimize the spectrum usage as a whole (in contrast to previous

© Springer International Publishing AG 2016
T.K. Madsen et al. (Eds.): MACOM 2016, LNCS 10121, pp. 57–71, 2016.
DOI: 10.1007/978-3-319-51376-8_5

standard amendments, such as 802.11n and 802.11ac, focused on increasing of the throughput at each link).

Different mechanisms are included in the current 802.11ax draft for improving the spectrum usage; these mechanisms work at the physical layer, exploiting multi-user (MU) MIMO and OFDMA, as well as at the MAC layer. While these solutions are more consolidated because they have been adopted by other technologies, interference mitigating solutions are still under debate, because there is no final demonstration of their effectiveness in generic scenarios.

A promising way to address this problem is to study how interference is mitigated in multi-hop wireless networks which have attracted much interest during last decades. For example, defining Wi-Fi Mesh technology, IEEE 802.11s introduces a mechanism called Mesh Deterministic Access (MDA or MCCA), which allows a station (STA) to reserve in advance a series of time intervals when only this STA can transmit, while all other STAs in the neighborhood of both the sender and receiver are forbidden to access the channel. To protect transmission, STAs periodically disseminate information about reservations among neighborhood. In order to minimize advertisement overhead, the time intervals are periodic and of the same duration, so the whole series of intervals is described with just three parameters specifying interval duration, period, and position of the first interval with respect to a beacon. The advantages of this scheme are so evident, that it is used for deterministic access in mm-Wave Wi-Fi (802.11ad) [7] and for periodic restricted access window in Wi-Fi for Internet of Things (802.11ah) [8]. Similar approach is also used in the IEEE 802.11aa networks to coordinate schedules of contention-free HCCA transmissions in overlapped hotspots, which is known as HCCA TXOP Negotiation. However, to use HCCA, exact traffic requirements like bandwidth, maximal loss rate, maximal delay shall be known by the Wi-Fi devices. Such limitation prevents HCCA (and consequently HCCA TXOP Negotiation) from wide usage.

Another way to improve spatial reuse is managing the condition which determines when the medium is considered to be idle. In Wi-Fi networks, to understand if the medium is idle, a STA performs carrier sensing. Basically, it measures signal+noise level in the channel. If the measured value is less than some *constant* sensitivity threshold, then the medium is idle and can be used for the transmission, otherwise it is considered to be busy. In case of dense deployment scenario, the interference from overlapping networks (or Overlapped BSSs, OBSS, in terms of 802.11) can result in high signal+noise level sensed by the STAs and thus can almost always block their transmissions. In this case, proper increase of the sensitivity threshold value allows the STAs to transmit without destructing alien transmissions. Such an approach, known as Dynamic Sensitivity Threshold (DSC) [9,10], is considered as a way to improve spatial reuse by 802.11ax developers.

Obviously, TDMA and DSC are opposite solutions. While TDMA tries to eliminate hidden node effect, DSC is designed to avoid exposed node effect. In the paper, we show that neither TDMA nor DSC alone can significantly improve performance in dense networks, while together they synergy. We also present a novel approach of joint usage of TDMA and DSC to flexibly divide channel time between OBSSs in rather complex scenarios. While DSC reduces the hotspot coverage, TDMA adaptively reduces the number of interfering networks and makes their transmissions orthogonal, by allowing each overlapping BSS to access the medium in a dedicated time interval. Unlike MCCA, where channel time reserved by a STA is wasted if the STA has no data for transmission, in our approach time intervals are adaptively assigned to a BSS or a subset of STAs, which contend for the channel. This provides better channel utilization and allows avoiding long guard intervals.

Although a number of papers study performance of pure CSMA/CA [11], TDMA [12,13] or DSC [9,10,14] in dense networks, to our best knowledge, this is the first paper which brings together these two opposite concepts—TDMA and DSC—and studies how fruitfully they can work together.

The rest of the paper is organized as follows: in Sect. 2, we briefly present the current 802.11ax draft. Section 3 presents preliminary results for the low-scale scenario with two interfering BSSs, which allows us dissecting the different effects of the interference on the 802.11 CSMA-based protocol. We present our solution in Sect. 4 and evaluate its performance in large-scale scenario in Sect. 5. Implementation remarks and testbed results are given in Sect. 6. Finally, conclusions are drawn in Sect. 7.

2   The IEEE 802.11ax Amendment

To improve performance in dense networks, 802.11ax introduces revolutionary features in addition to new modulation and coding schemes (MCSs).

First, 802.11ax networks use OFDMA for MU transmission both in uplink (UL) and downlink (DL). Although the legacy OFDM is designed for wide channels, it still suffers from frequency selective interference and does not allow efficient use of 160 MHz channels enabled by 802.11ac. Therefore dense hot-spots fall back to 20 MHz channels with 8 times lower throughput. In contrast, with OFDMA, the sender can use only the best resource unit (RU), i.e. set of tones, while other RUs can be used by other STAs. 802.11ax provides a flexible mechanism of splitting the channel into RUs. Specifically, each 20 MHz subchannel can be split into up to 9 RUs.

Second, 802.11ax enables UL MU-MIMO transmission. Both MU-MIMO and OFDMA can be jointly used. The main limitation is connected with the duration of UL transmissions, which shall be the same for all simultaneously transmitting STAs. Both MU-MIMO and OFDMA require tight synchronization, which is provided in the following way. Each UL MU transmission follows a Trigger frame sent by an AP. This frame allocates RUs for particular STAs and defines duration, MCSs and other parameters for the UL transmission. An important feature of the novel amendment is that some RUs can be allocated for Aloha-like random access. Such a random access is favorable for transmitting short sporadic frames or for notifying the AP about non-empty queues.

Third, the 802.11 Working Group continues moving MAC signaling to the PHY header (which is widely used in 802.11ah). Specifically, in the 802.11ax PHY header, a transmitting STA can indicate the remaining duration of TXOP, network color (a randomly selected short network identifier introduced in 802.11ah), transmission direction (UL or DL) and some other information. Such a modification improves many aspects of network operations: neighboring STAs can obtain TXOP duration even if they do not receive correctly the whole frame; a STA can switch off its radio for the frame duration if it starts receiving an UL frame originated from the same network; a STA can consider the medium to be idle, if the frame with an alien color is received with very low power.

Fourth, 802.11ax improves power efficiency by importing and extending Target Wakeup Interval, a beacon-free power saving mechanism developed in IEEE 802.11ah. This allows devices to sleep for long time except for predefined intervals, without the need for periodic awakenings to receive beacons.

An important issue, which should be considered by the group is *improving spatial reuse and avoiding hard collisions* inherent in dense deployment of BSSs. However by today, the results are not so bright as expected, since the task group cannot reach a consensus on how exactly such mechanisms shall work. The group agreed to distinguish virtual carrier sense (NAV) for intra-BSS and inter-BSS transmissions. They agreed to have different energy detection and carrier sense thresholds (determining when the channel is busy) for inter-BSS and intra-BSS transmission; finally, they agreed to have flexible power control, which allows per-frame changing transmission power in order to reduce the frame-induced interference at other STAs. Many solutions presented in the group have not been approved, either because of the lack of thorough performance evaluation or manufacturer business issues, e.g. roster [15], dynamic sensitivity control (DSC) [9,10], etc.

3   Preliminary Analysis of Inter-BSS Interference

To show the dramatic performance degradation caused by inter-BSS interference, we consider a scenario similar to that from [17] and perform simulation with ns-3 [18]. We consider two BSSs each of which consists of an AP and a client STA, see Fig. 1. Each AP has saturated UDP traffic destined to the associated STA. The main simulation parameters are summarized in Table 1. We run simulation with both 802.11a and 802.11ac PHYs. However the effects are similar, so we present only results for 802.11ac. In our study, we vary the distance d between two APs and consider four cases: with/without RTS/CTS and with/without aggregation. The results shown in Fig. 2a–d were obtained for the fixed MCS 5, and depend on the following key distances. $R_{CCA} = 50\,\text{m}$ is the minimal distance between two nodes at which they cannot sense transmission from each other. $R_{MCS0} = 49\,\text{m}$ and $R_{MCS5} = 27\,\text{m}$ are the maximal distances between transmitter and receiver at which frames sent with MCS 0 and MCS 5 respectively, are correctly received with probability close to 1. $R_{MCS0}^{kill} = 20\,\text{m}$ and $R_{MCS5}^{kill} = 60\,\text{m}$ are the maximal distances from the receiver, at which an interfering node can kill an ongoing transmission at MCS 0 and MCS 5, respectively. Given these values, we can define the following inter-BSS interference areas.

Fig. 1. Network topology used for simulation in ns-3

Table 1. Main simulation parameters

Parameter	Value
PHY	VHT @80 MHz
AP/STA TX power	15 dBm
CCA threshold	−76 dBm
MCS for data/control frames	MCS 5/MCS 0
Path loss model	Defined in [16]
Channel access	EDCA, AC_BE
CW_{min}, CW_{max}	15, 1023
MPDU size	1538 Bytes
Aggregation/BlockACK	32 MPDUs
Retry limit	7

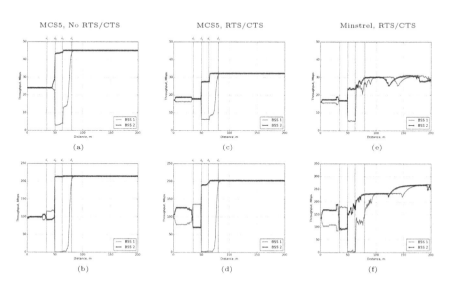

Fig. 2. Numerical results obtained with ns-3: without aggregation/BlockACK (above) and with aggregation/BlockACK (below), without RTS/CTS (a–b) and with RTS/CTS (c–f)

Soft interference. When $0 \leq d < d_1 = 35$m, all the nodes are within R_{CCA} range and the spectrum should be fairly shared. However for RTS/CTS, we get unfairness: when RTS frames are simultaneously sent, STA1 cannot receive AP1's RTS, whereas STA2 correctly receives RTS from AP2, since the distance between AP1 and STA2 is longer than R_{MCS0}^{kill}.

When $d_1 \leq d < d_2 = 50$ m, AP1 and STA2 become out of R_{CCA} range. Since the distance between AP1 and AP2 is greater than R_{MCS5} they cannot decode transmission of each other, therefore after AP2 transmits, AP1 waits for EIFS=$96\mu s$ and then contends for the channel, while AP2 waits for an acknowledgment. Without aggregation, an ACK is transmitted; so both APs start channel contention simultaneously. However, in case of aggregation, STA2 transmits BlockACK which is longer than normal ACK ($68\mu s$ vs $44\mu s$). Although EIFS is enough to protect BlockACK from interference, AP2 starts contending for the channel later than AP1.

An interesting effect occurs if 45 m $\leq d < 50$ m at Fig. 2a. If the APs transmit simultaneously, the transmission of AP1 is damaged, because the distance between STA1 and AP2 is less than R_{MCS5}^{kill}, while STA2 successfully receives the packet. Thus BSS2 throughput is greater than BSS1 one.

Hard interference. When $d_3 \leq d < d_4 = 75$ m, the APs do not sense each other and severe hidden node effects occur. In this case, the transmission of AP2 is always successful because the distance between AP1 and STA2 is greater than R_{MCS5}^{kill}. However, since the distance between AP2 and STA1 is lower than R_{MCS5}^{kill}, AP2 kills AP1's transmission. The only opportunity for AP1 to succeed is to transmit in gaps when AP2 is counting down backoff, which is possible because of high MCS. Using short RTS helps to transmit *inside* such gaps, which results in higher throughput at Fig. 2c than in Fig. 2a. At the same time, aggregation— widely used in high-throughput Wi-Fi networks— aggravates unfairness for two reasons. First, AP1's frame transmission becomes so long that even if the first data frame is transmitted inside the gap, the following ones and the BlockACK can cross gap boundaries. Second, AP2's transmissions become longer, so the number of gaps per time interval decreases. When $d > R_{MCS5}^{kill}$, acknowledgments transmitted by STA2 cannot damage transmission of AP1, so the gaps available for AP1's successful transmissions become longer and the throughput of BSS1 increases.

No interference. Finally, when $d \geq d_4$, the interference between BSSs becomes negligibly low, since the distance among nodes of different BSSs is more than R_{MCS5}^{kill}, which gives maximal throughput for both BSSs.

As shown in Fig. 2c–d, RTS/CTS cannot significantly improve the performance in case of hidden nodes, though providing a bit better results in area $d_3 \leq d < d_4$ for BSS1.

We have also studied interference with joint usage of RTS/CTS and adaptive MCS, considering the Minstrel rate control algorithm [19], which tries to use such a rate that taking into account loss ratio gives the highest throughput. As shown in Fig. 2e,f, the results are rather similar to those with fixed MCS, therefore adaptive rate mechanisms do not mitigate hidden node effect.

4 Solution Design

As it was mentioned in Sect. 1, DSC is a promising solution for enhancing network performance in case of OBSSs. The key idea of this approach is increasing sensitivity threshold value at APs and STAs and thus prevent blocking their transmissions by OBSS transmissions. The question is how to choose the value for sensitivity threshold refered to as DSC Threshold in this case. Obviously this value shall allow STAs to receive transmissions within their own BSS. Thus DSC Threshold shall be less than $TxPower - \max_{i,j} PathLoss(i,j)$, where $TxPower$ is the transmission power which is considered to be the same for the AP and all the STAs within the BSS for simplicity, and $PathLoss(i,j)$ is signal attenuation between STA_i and STA_j within the considered BSS. In papers considering DSC [9,10], it is usually proposed to set DSC Threshold to $TxPower - PathLoss - margin$, where $PathLoss$ is propagation losses between an AP and the farthest associated STA in the BSS, and margin is a positive value between 18 and 25dB. Note that in real life the DSC Threshold shall not exceed the maximum value defined in regulatory documents (e.g., ETSI EN 300 328 [20]). In further analyses we consider two values of DCS margin: 20 dB (which meet the original proposal) and 1 dB (which is close to zero).

Although usage of DSC allows increasing the overall network throughput in dense deployment, there can be situations when even with DCS the STAs of different BSSs sense transmission of each other. Moreover, as widely shown in literature, e.g. [9,10], by eliminating exposed node effect, DCS opens the door for the hidden node effect.

In such a situation, the best solution is to make transmissions in overlapping BSSs orthogonal. Unfortunately, the limited number of Wi-Fi channels does not always allow implementing a traditional frequency planning among overlapping BSSs. Another simple mechanism for making the transmissions performed in different BSSs orthogonal can be TDMA. Time-based allocations are already considered in the recent standard amendments (e.g. 802.11s, 802.11ad), where a STA—the owner of the reserved time interval—can keep the channel by performing multiple frame transmissions. However, while legacy reserved intervals are allocated to a single STA, our idea is the generalization of this mechanism assigning reserved intervals to a subset of STAs or the entire BSS and allowing to access the channel according to the legacy contention rules.

Of course, if BSSs can only transmit within the reserved time intervals, we might have some performance loss because the two APs could simultaneously transmit to stations that are not in their overlapping areas. Thus, channel reservation made by a STA may have negative effects on the throughput of other STAs. That is why we propose to establish reservation only if an AP somehow understands that the hidden node effect prevents it from delivering data to some STAs. Specifically, because of the FIFO policy, if the AP experiences hard interference for at least one link (see Sect. 3), the throughput for all downlinks of this AP will be close to zero. In this case, we propose the AP to reserve some portion of channel time for future transmissions. While the detail description of the mechanism, how to reserve channel time and advertise reservations among neighboring APs is left for future study, in Sect. 5 we evaluate the concept of joint DSC and TDMA operation, assuming that we have some wired or wireless signaling between APs, and the APs can reserve periodic sequences of time intervals with duration of 20 ms.

5   Simulation Results

In order to evaluate performance of the joint usage of DSC and TDMA, we use ns-3 and design a scenario typical for airports, mass events, malls, etc. In this scenario, we have a hexagonal grid of 12×18 APs operating on one of three channels. The grid step is d. The channels are chosen in such a way that the distance between APs operating on the same channel is maximized, see Fig. 3. To model irregularity typical for real deployments, we choose the position of each AP randomly in $\frac{d}{4}$ neighborhood of the grid vertexes. In the scenario, we have $12 \times 18 \times 3 = 648$ STAs uniformly distributed among the Internet access area. Each STA is associated with the closest AP, which transmits saturated traffic to it using Minstrel rate control algorithm. To speed up simulation of such a large network, we consider 802.11a PHY and 20 MHz channels. Other MAC and PHY simulation parameters are summarized in Table 2.

Table 2. Main simulation parameters

Parameter	Value
AP/STA TX power	15 dBm
CCA threshold	−91 dBm/DSC threshold
Rate control	Minstrel
MPDU size	1538 Bytes
Retry limit	7

In the experiments, we vary grid step d and measure throughput for each STA. We consider eight cases: default CCA Threshold or DSC, with or without TDMA reservations, with or without RTS/CTS. Since the results for enabled and disabled RTS/CTS handshake are quite similar, here we present only the results for RTS/CTS switched off.

When neither DSC nor TDMA is used, the average throughput reaches a peak of 3.2Mbps when $d = 145$. Such a result is caused by two opposite effects. First, the higher is d, the better is spatial reuse, and the less is contention for the channel, which results in average throughput increase. Second, when d increases, the average distance between the AP and the STAs increases and the AP selects lower and lower rate for transmission. Finally, when d exceeds 100 there can be situations, when data cannot be delivered to some STAs too poor signal strength. These effects perfectly explain the results, presented in Fig. 4a.

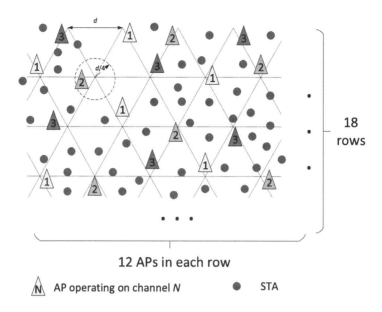

Fig. 3. Network topology

Specifically low average throughput for low d is explained by the first effect, while almost zero minimal throughput at $d > 100m$ is explained by the second one. The throughput fluctuation for $25 < d < 100$ is caused by both the hidden node effect and exposed node effect.

As expected, switching on DSC, as described in Sect. 4 significantly increases the average throughput for all distances, but opens a door for starvation caused by the hidden node effect. It results in zero minimal throughput for several BSSs, which means that DSC can block the whole BSS, see Fig. 4. At the first sight, the situation can be improved by using large margin, which plays a guard role and should protect from hidden STAs. However, as shown in Fig. 4, DSC with large margin does not solve starvation problem, but decreases the average throughput.

Obviously, TDMA alone cannot improve the average throughput, especially in case of dense networks where the channel shall be divided between multiple BSSs. However it makes the channel resource allocation more fair. Thus the joint usage of DSC and TDMA improves both the average and minimal throughput for the large area of d, as shown in Fig. 4. Specifically, except for too low $d \le$ 5m—which corresponds to very dense deployment and high contention—and $d > 100m$—where data cannot be delivered for some STAs at all—the maximal throughput and the minimal one are much higher than without DSC and TDMA.

Thus, we should not debate whether to use either DSC or TDMA in the new generation Wi-Fi, but should combine these techniques to achieve the best performance.

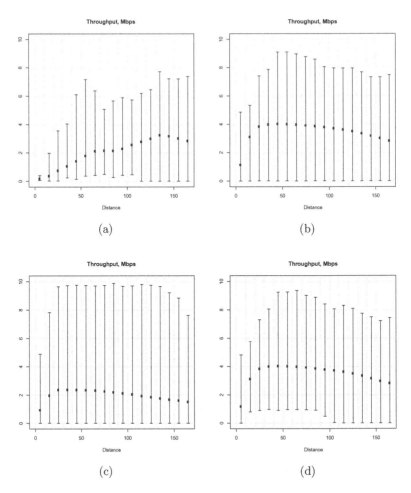

Fig. 4. Minimal, maximal and average throughput vs grid step d without RTS/CTS: (a) No TDMA, no DSC; (b) No TDMA, DSC with $margin = 1\,dB$; (c) No TDMA, DSC with $margin = 20\,dB$; (d) TDMA, DSC with $margin = 1\,dB$.

6   Experimental Validation

In order to implement the proposed approach in real equipment and experimentally validate it, it is required to support the inter-BSS synchronization, DSC and the allocation of per-BSS reserved intervals. In this Section, we mostly focus on inter-BSS synchronization and TDMA, which include time critical operations, that cannot be managed at the driver level. Modifications of such MAC operations are usually addressed by using SDR platforms (having limitations on the PHY) or card firmware (usually unavailable). In both cases, the implementation is time consuming, because regardless of the complexity of the function to be modified, it is necessary to study the hardware-specific architecture and deal with low-level programming models.

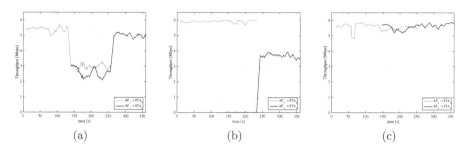

Fig. 5. Throughput under different inter-BSS interference conditions: soft (a), hard (b), and no interference (c).

In our work, we have succeeded in implementing the inter-BSS TDMA scheme in few days, by exploiting the so called Wireless MAC Processor (WMP) architecture [21] prototyped on top of a commercial card by Broadcom. Rather than implementing a specific protocol, our WMP firmware implements a generic executor of state machines and exposes some elementary hardware signals (e.g. start of a frame reception, expiration of a timer, sensing of a channel busy interval, etc.) and actions (transmit a frame, set a timer, switch to sleep mode, etc.) to experimenters. Different MAC protocols can then be loaded into the card in terms of programmable state machines. The platform has been proved to be flexible enough to run different variants of CSMA and TDMA-based protocols.

Starting from the state machine implementing an intra-BSS TDMA protocol and exploiting the platform graphical editor, we easily defined a novel state machine for the proposed scheme. Specifically, we added the inter-BSS synchronization mechanism, by specifying an action at the AP side for correcting the local TSF upon the reception the new information about channel reservations. This common reference signal is then used by the TDMA state machine for setting the slots in the BSS reserved interval in which the STAs are allowed to access the channel. The interval allocation within the frame is a configuration parameter, that has been set to different values for the two BSSs (odd slots to AP_1 and even slots to AP_2). The interval size is also a tunable parameter, which has been set to the minimum interval required for transmitting a frame of 1500 bytes at 11 Mbps, in order to test our system under tight synchronization requirements. Figure 6 demonstrates the efficiency of our TDMA implementation (with/without inter-BSS synchronization) by showing channel occupancy in terms of power levels measured by a USRP platform used as a PHY sniffer. Different power levels and occupancy intervals allow to easily recognize the long data frames sent by APs and the short ACKs sent by STAs.

Since running large-scale experiments similar to those described in Sect. 5 is hardly possible with a testbed, for testbed experiments we deployed a scenario with two overlapping BSSs, see Sect. 3.

Figure 5 summarizes the throughput results obtained under decreasing levels (from left to right) of inter-BSS interference, without RTS/CTS and for a data

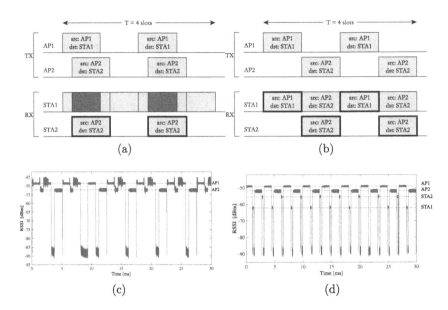

Fig. 6. Unsynchronized TDMAs: STA_1 receives two frames and STA_2 none (a). Synchronized TDMAs: STAs receive two frames each (b). Real trace with unsynchronized (c) and synchronized TDMAs (d).

rate set to 11 Mbps. All experiments are organized into three phases lasting 120s: first, only one UDP flow from AP_2 is active, then the two APs run their flows in parallel, finally only AP_1 flow remains active. In Fig. 5a, both the APs transmit at the maximum transmission power of 20dBm and can perfectly hear each other; in these conditions, they basically share the channel capacity and get one half of the total throughput each, as expected in the *soft interference area*. The phenomenon gets worse in Fig. 5b, corresponding to the *hard interference area* when the transmission power is reduced to 1dBm and the two APs consistently fail in sensing each other. Finally, in Fig. 5c corresponding to the *no interference area*, we used a 6dB attenuator for further reducing the inter-BSS interference and we observe that the two BSSs work as in isolated conditions, because they do not interfere with each other.

We finally evaluate the efficiency of our inter-BSS TDMA scheme in all the scenarios described above, by also considering static and dynamic assignments of intervals to each BSS. In the static scenario, we consider a periodic frame of two intervals and we assign the odd ones to BSS_1 and the even ones to BSS_2. Obviously, this static assignment is not optimal when only one traffic flow exists. In the dynamic scenario, we assume that the assignments of reserved intervals can vary over time, according to the load experienced in each BSS. When only BSS_1 or BSS_2 is active, both reserved intervals of the frame are assigned to the active BSS; when both the BSSs are active, each BSS receives only one assignment according to the usual odd/even schedule. Figure 7 shows the throughput results

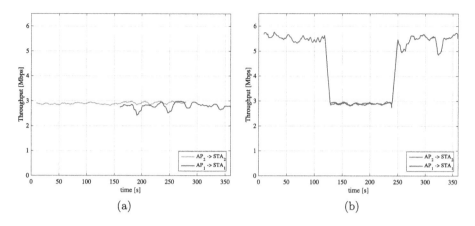

Fig. 7. Throughput for static (a) and dynamic (b) TDMA.

obtained under inter-BSS TDMA, in the static and dynamic case. Results do not depend on the interference areas, because the adoption of time-division makes the two BSSs completely orthogonal.

7   Conclusions

In this paper, we have studied interference in dense deployment, which is a hot scenario for both Wi-Fi operators and standard developers. We proved that the hidden node effect is inherent in such scenarios and it occurs even if the distance between an AP and associated STAs is small. Then we showed that neither RTS/CTS nor adaptive rate control algorithms can overcome hidden node effect, which results in almost zero throughput for blocked STAs. Thus the problem of hidden nodes shall be taken into account by the developers of 802.11ax, which aims to improve efficiency in dense networks.

To provide fairness and avoid performance degradation, we have proposed to jointly use DSC and TDMA. As for TDMA, it mainly inherits the standardized channel reservation protocols. However, in contrast to existing Wi-Fi reservation-based protocols, e.g. MCCA, in our scheme the channel can be reserved not only for a particular STA but also for a subset of STAs or the whole BSS, which improves channel reuse if the STA has no traffic. Experimental results show high efficiency of the proposed scheme and we intend to contribute to 802.11ax.

While in the present work we did not consider any signaling between the APs, and the TDMA structure was fixed, in future work, we are going to develop a method, which adaptively selects parameters for channel reservations. Apart from that we are going to evaluate performance of the proposed solution in a scenario with OFDMA transmissions and different contention parameters, which was recently enabled in 802.11ax [22].

Acknowledgment. This work has been partially funded by the Russian Science Foundation, agreement No 16-19-10687 (Sects. 3–5), and by the WiSHFUL EU H2020 Project, GA No. 645274 (Sects. 4–6).

References

1. Ergin, M.A., Ramachandran, K., Gruteser, M.: An experimental study of inter-cell interference effects on system performance in unplanned wireless LAN deployments. Comput. Netw. **52**(14), 2728–2744 (2008)
2. Papagiannaki, K., Yarvis, M., Conner, W.: Experimental characterization of home wireless networks and design implications. In: Proceedings of the 25th IEEE International Conference on Computer Communications, INFOCOM 2006, pp. 1–13, April 2006
3. Bicket, J., Aguayo, D., Biswas, S., Morris, R.: Architecture, evaluation of an unplanned 802.11b mesh network. In: Proceedings of the 11th Annual International Conference on Mobile Computing and Networking, ser. MobiCom 2005, pp. 31–42. ACM, New York (2011)
4. Jardosh, A.P., Mittal, K., Ramachandran, K.N., Belding, E.M., Almeroth, K.C.: IQU: practical queue-based user association management for WLANs. In: In: Proceedings of MobiCom, pp. 158–169 (2006)
5. Tinnirello, I., Scalia, L., Campoccia, F.: Improving IEEE 802.11 performance in chain topologies through distributed polling and network coding. In: 2009 IEEE International Conference on Communications, pp. 1–6, June 2009
6. Khorov, E., Kiryanov, A., Lyakhov, A.: IEEE 802.11ax: how to build high efficiency WLANs. In: 2015 International Conference on Engineering and Telecommunication (EnT), pp. 14–19, November 2015
7. Khorov, E., Lyakhov, A., Ivanov, A., Zankin, V.: Modelling deterministic channel access in millimetre wave Wi-Fi. In: Proceedings of 12th International Symposium on Wireless Communication Systems (ISWCS 2015) (2015)
8. Khorov, E., Lyakhov, A., Krotov, A., Guschin, A.: A survey on IEEE 802.11 ah: an enabling networking technology for smart cities. Comput. Commun. **58**(C), 53–69. http://dx.doi.org/10.1016/j.comcom.2014.08.008
9. Afaqui, M.S., Garcia-Villegas, E., Lopez-Aguilera, E., Smith, G., Camps, D.: Evaluation of dynamic sensitivity control algorithm for IEEE 802.11ax. In: 2015 IEEE Wireless Communications and Networking Conference (WCNC), pp. 1060–1065, March 2015
10. Afaqui, M.S., Garcia-Villegas, E., Lopez-Aguilera, E., Camps, D.: Dynamic sensitivity control of access points for IEEE 802.11ax. In: IEEE International Conference on Communications (ICC 2016), May 2016
11. Bellalta, B., Zocca, A., Cano, C., Checco, A., Barcelo, J., Vinel, A.: Throughput Analysis in CSMA/CA Networks Using Continuous Time Markov Networks: A Tutorial. Springer International Publishing, Cham (2014)
12. Zheng, L., Hoang, D.B.: Performance analysis for resource coordination in a high-density wireless environment. In: IEEE Symposium on Computers and Communications, pp. 685–690. IEEE (2008)
13. Tinnirello, I., Gallo, P.: Supporting a pseudo-TDMA access scheme in mesh wireless networks. In: Bianchi, G., Lyakhov, A., Khorov, E. (eds.) WiFlex 2013. LNCS, vol. 8072, pp. 80–92. Springer, Heidelberg (2013). doi:10.1007/978-3-642-39805-6_8

14. Mvulla, J., Park, E.-C., Adnan, M., Son, J.-H.: Analysis of asymmetric hidden node problem in IEEE 802.11 ax heterogeneous WLANs. In: 2015 International Conference on Information and Communication Technology Convergence (ICTC), pp. 539–544. IEEE (2015)

15. Coffey, S.: Airtime analysis of EDCA. Technical report (2015). https://mentor. ieee.org/802.11/dcn/15/11-15-1114-01-00ax-airtime-analysis-of-edca.pptx

16. Merlin, S.: TGax Simulation Scenarios (2014). https://mentor.ieee.org/802.11/ dcn/14/11-14-0980-16-00ax-simulation-scenarios.docx

17. Krasilov, A.: Physical model based interference classification and analysis. In: Vinel, A., Bellalta, B., Sacchi, C., Lyakhov, A., Telek, M., Oliver, M. (eds.) MACOM 2010. LNCS, vol. 6235, pp. 1–12. Springer, Heidelberg (2010). doi:10. 1007/978-3-642-15428-7_1

18. The ns-3 network simulator. http://www.nsnam.org/

19. Xia, D., Hart, J., Fu, Q.: On the performance of rate control algorithm Minstrel. In: 2012 IEEE 23rd International Symposium on Personal, Indoor and Mobile Radio Communications - (PIMRC), pp. 406–412, September 2012

20. ETSI EN 300 328 V1.9.1. Electromagnetic compatibility and Radio spectrum Matters (ERM); Wideband transmission systems; Data transmission equipment operating in the 2,4 GHz ISM band and using wide band modulation techniques; Harmonized EN covering the essential requirements of article 3.2 of the RTTE Directive

21. Tinnirello, I., Bianchi, G., Gallo, P., Garlisi, D., Giuliano, F., Gringoli, F.: Wireless MAC processors: programming MAC protocols on commodity hardware. In: 2012 Proceedings of IEEE INFOCOM, pp. 1269–1277. IEEE (2012)

22. Khorov, E., Loginov, V., Lyakhov, A.: Several EDCA parameter sets for improving channel access in IEEE 802.11ax networks. In: International Symposium on Wireless Communication Systems (2010)

Generic Energy Evaluation Methodology for Machine Type Communication

Thomas Jacobsen[1(✉)], István Z. Kovács[2], Mads Lauridsen[1], Li Hongchao[3],
Preben Mogensen[1,2], and Tatiana Madsen[1]

[1] Department of Electronic Systems, Aalborg University, Aalborg, Denmark
tj@es.aau.dk
[2] Nokia Bell Labs, Aalborg, Denmark
[3] Nokia, Beijing, China

Abstract. It is commonly accepted that the 3rd Generation Public Partnership Long Term Evolution (also known as 3GPP LTE) standard is likely to be unfit for future large scale machine type communication (MMTC). As a result, a new standard, LTE Narrow-Band Internet of Things (NB-IoT) and several radio protocol proposals are being developed. One of the main performance indicators for MMTC is the radio energy consumption. It is important to be able to evaluate the energy consumption of the new standard and the proposed protocols, therefore a generic energy consumption evaluation methodology tailored for MMTC devices is required. Such methodology is the contribution of this paper. It is developed by defining a generic radio transmission and describing the factors which affect the energy consumption. Special attention is put on the factors; power control, link-level performance and a radio power model with a non-constant power amplifier (PA) efficiency model intended for MMTC devices. The results show the impact of the factors and highlight first that applying a commonly used constant radio PA efficiency model can result in an overestimation of the battery life of up to 100% depending on the traffic scenario. It is also highlighted that combining power control, transmit repetitions and the radio power model opens for new methods to minimize the radio energy consumption.

1 Introduction

The 3rd Generation Public Partnership (3GPP) Long Term Evolution (LTE) can become unfit for large scale (massive) machine type communication (MMTC) [8,14,16]. An example is the LTE access procedure which is known to become congested when serving a massive number of devices [13]. As a consequence new standards are being developed, like 3GPP Narrow-Band Internet of Things (NB-IoT) and LTE for MTC (LTE-M) [1–3]. In parallel with the standardization work on NB-IoT and LTE-M, there was and still is, significant research ongoing regarding MMTC protocols, as reported in [9,11,13,15]. As it is important to be able to evaluate the new standards and the proposed protocols, a generic methodology is required which is applicable for all proposed protocols and new standards.

© Springer International Publishing AG 2016
T.K. Madsen et al. (Eds.): MACOM 2016, LNCS 10121, pp. 72–85, 2016.
DOI: 10.1007/978-3-319-51376-8_6

MMTC is generally characterized as communication with infrequent small payloads, in scenarios with high device density. The devices can be in challenging coverage conditions and have extreme battery life requirements. Therefore one of the main performance indicators for MMTC is the radio energy consumption [13].

The most common energy evaluation methodology is given in [3] and targets NB-IoT which is the state-of-the-art standard for MMTC. The common energy evaluation methodology does not include the energy consumption impact from challenging coverage conditions or the interference caused by a high density of devices. Neither does it include a realistic model of the power amplifier (PA) energy efficiency. The authors of [12] show that assuming a constant PA efficiency is not valid for smartphones from 2013–2014. Even though older smartphone radios cannot be directly compared to the radio in MMTC devices, it seems unlikely that the efficiency of an MMTC radio PA will be constant as it is assumed in [3].

This paper presents a generic energy evaluation methodology tailored for MMTC. The methodology includes important features for MMTC, such as uplink power control to manage the level of interference occurring from a high density of devices, transmit repetitions to cope with challenging coverage conditions, and a radio power model intended for MMTC devices. The methodology is generic through its model of a transmission. The power model is based on [3], where we propose to use a non-constant PA efficiency model intended for MMTC radios, derived from empirical measurements on smartphones [12].

The paper is organized as follows. Section 2 presents the generic energy evaluation methodology along with the revised radio power model. In Sect. 3 we apply the proposed methodology and demonstrate the impact of its input. The results and implications of the model is discussed in Sect. 4. The paper is concluded in Sect. 5 which also outlines the future work.

2   Generic Energy Evaluation Methodology

For the energy evaluation methodology to be applicable to the new standards and proposed protocols it needs to be generic. To achieve this we have identified the most important factors that affect the energy usage in a radio transmission. The identified factors are illustrated in Fig. 1. These are channel aspects such as radio fading and interference, power control, link-level performance, power model and radio access configurations such as transmit repetitions. Modelling of specific protocols which utilize several radio transmissions can be done as a chain of radio transmission blocks. The following sections will describe these factors.

2.1   Radio Fading and Interference

MMTC devices can experience challenging radio coverage conditions [1–3] e.g. due to being located deep indoors.

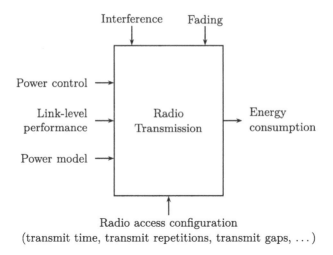

Fig. 1. Characterization of the input and outputs of a radio transmission. The inputs are power control, link-level performance, power model and the radio access configurations. The transmission is affected by interference and fading. The output is the radio energy consumption

One method to overcome the effect of a large path loss is by repeating transmissions in time [3]. When doing so, the receiver combines the received transmissions to increase the energy of the desired signal. The configuration of transmit repetitions is a part of the access configuration which also dictates when and how often to transmit. The number of transmit repetitions needs to be taken into account by the power control to manage the level of interference. With the use of repetitions the quantity of devices active in a single time slot (TTI) depends on how many devices start their transmission and how many are already repeating their transmissions.

2.2   Power Control

In order to control the level of interference, power control is included in the methodology and for simplicity reasons open loop power control (OLPC) is chosen. OLPC aims to equalize the signal strength from the devices at the base station (BS) receiver. In the OLPC implementation, used in this energy evaluation methodology, it is chosen to use the number of simultaneous transmitting devices as the traffic intensity (M) and the path loss compensation factor (α) along with the target received signal strength at the BS (P_0) as the information which is broadcast to the transmitting devices. The number of active devices can be estimated by the BS through e.g. multi-user-detection techniques. The details on how this is done is out of the scope of this paper.

Each device uses the broadcast information to calculate which transmit power (Ptx) they should use. The transmit power is calculated using (1), where Ptx_{dBm} and $Ptx_{max,\text{dBm}}$ is the transmit power and maximum transmit power in dBm and PL_{dB} is the path loss in dB. The devices estimate the path loss from the downlink reference received signal strength.

$$Ptx_{dBm} = min(Ptx_{max,dBm}, P_{0,dBm} - PL_{dB} \cdot \alpha) \quad [dBm] \qquad (1)$$

The target received power at the BS (P_0) is determined from the desired signal to interference and noise ratio (SINR) which dictates the target performance of the transmission. The target SINR is derived from link-level performance curves, simply referred to as performance curves in this paper and are described in further detail in Sect. 2.3. The target SINR (γ_{SINR}) can be expressed as in (2). The interference is caused by all other transmitting devices ($M - 1$). The noise power is denoted by N. Notice that (2) uses the linear version of PL_{dBm} and Ptx_{dBm} and implies that all devices use the same number of transmit repetitions (R) and that it is only valid when $R > 0$, $Ptx \leq Ptx_{max}$ and as long as the path loss can be compensated.

$$\gamma_{SINR} = \frac{(Ptx \cdot PL^{\alpha}) \cdot R}{(M - 1) \cdot (Ptx \cdot PL^{\alpha}) \cdot R + N} \quad [1] \qquad (2)$$

2.3   Link-Level Performance Curves

The SINR affects the receiver's probability of correctly decoding the transmitted message. By setting a target performance (e.g. 10% successful decode probability) the corresponding target SINR (γ_{SINR}) can be found. The use of performance curves (successful decoding probability vs SINR) enable the evaluation methodology to support any coding and modulation, data type and multiple access technique with different multi user detection abilities. The corresponding SINR can be translated to SNR by considering the interference as noise.

An example of two performance curves (denoted curve A and B) are shown in Fig. 2. They are generated by a link level simulator, which has been simulating LTE PRACH sequences [4] (Zadoff-Chu sequences) of two different lengths. The performance shows the probability of the eNB not successful decoding the PRACH sequence at various SNR. In the figure the target performance is set to 10% error rate which translates to a target SINR (γ_{SINR}) of -18.9 dB and -11.0 dB for curve A and B respectively. The better SINR performance of A comes at the cost of taking more time to transmit. In this example, A requires 0.8 ms (without cyclic prefix which takes 0.103 ms) and B requires 0.134 ms to transmit. The pair of a performance curve and transmit time is denoted a mode through the rest of the paper.

2.4   Power Model

The radio power model used in this evaluation methodology originates from [3]. The model is updated with a PA efficiency model which is based on the work presented in [12] and modified to be used for MMTC radios. The radio power model from [3] utilizes four power states; power saving mode (PSM), receiving (RX),

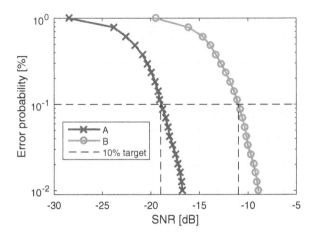

Fig. 2. Example performance curves A and B.

idle (*Idle*) and transmitting (*TX*). All states are included in the evaluation in order to include the energy impact of synchronization, configurations receptions, gaps between transmissions and receiving downlink traffic. A transmission with all four states (*PSM*, *RX*, *Idle*, and *TX*) is depicted in Fig. 3. Notice this is an example of what occurs in each state and the order of the states.

The transmission starts with the radio sleeping for a certain period before waking up from power saving mode (*PSM*). The time spent in the *PSM* state (T_{PSM}) depends on the traffic model, sleep configurations, and whether uplink data is ready for transmission.

Once the radio is awake it will change to *RX* state where it will start acquiring downlink synchronization such that it is able to decode the broadcast channel

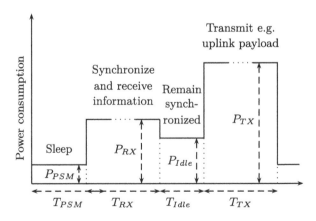

Fig. 3. Power model states (*PSM*, *RX*, *Idle* and *TX*) in a power and time domain with examples of what happens in each state.

(and control channel) to receive the configuration information. This information includes power control configurations, access configurations and scheduling
grants (if the protocol utilize scheduled access). If downlink data are scheduled
for the device, the radio will acquire the data in the RX state. The time spent
in the RX state therefore depends on whether downlink payload is available for
the device, the payload size, modulation and coding and the SINR. Devices in
bad coverage can be assumed to spend more time to acquire synchronization
compared to others with better coverage conditions. The power consumption in
the RX state (P_{RX}) is assumed to be independent of the modulation and coding
scheme, data rate and bandwidth which is a reasonable assumption according
to [12].

If the radio has uplink data to transmit, it will change to the TX state. The
time spent in the TX state (T_{TX}) depends on the configured number of transmit
repetitions (R), gaps between the transmit repetitions, uplink modulation and
coding scheme (UL MCS) and uplink payload size. The power drawn in the
transmit state (P_{TX}) depends on the transmit power dictated by the power
control and the efficiency of the radio PA which similarly depends on the transmit
power [12]. P_{TX} is similar to P_{RX} assumed to be independent of the modulation
and coding scheme, data rate and bandwidth.

Time spend on waiting (e.g. for an opportunity to transmit uplink payload
or in gaps between transmit repetitions) are spend in the $Idle$ state where the
radio maintains synchronization, as described in [3]. This is the main difference
between $Idle$ and PSM, where the radio in PSM is turned off such that synchronization cannot be maintained. The power draw in the PSM state (P_{PSM})
and the $Idle$ state (P_{Idle}) are device specific.

The energy consumption of a transmission can be calculated as the area below
the line in Fig. 3. The model proposed in this paper is given by (3). Notice that it
does not consider ramp up or ramp down time in state transitions as in [12].

$$E_{tot} = P_{PSM} \cdot T_{PSM} + P_{RX} \cdot T_{RX} \\ + P_{Idle} \cdot T_{Idle} + P_{TX}(Ptx_{\mathrm{dBm}}) \cdot T_{TX} \qquad \mathrm{[J]} \qquad (3)$$

The energy efficiency of the radio PA dictates the relation between the transmit power (Ptx) and the consumed power (P_{TX}). A common model from [3]
assumes that the energy efficiency is constant at either 30% or 40% independent
on the transmit power. A study conducted by [12] shows that this is not a valid
assumption for LTE smartphones in 2013–2014.

The research done in [12] might not be directly applicable in terms of absolute
power values in a power model intended for MMTC devices. Clearly the best fitting PA energy efficiency model would be derived from emperical measurements
from a MMTC device using NB-IoT or LTE-M radios. But as no such device
is commercially available (to the best of the authors' knowledge) the work presented in [12] is used to create the PA efficiency model intended for MMTC
radios.

One of the targets for NB-IoT and LTE-M is low cost [3]. If the PA should be
cheap and the bandwidth is lower (e.g. from 20 MHz in LTE to 1.4 MHz in LTE-M

or 200 kHz in NB-IoT), the PAs high-gain mode (described in [12]) used in smart-phones PAs might not be needed and hence can be removed. This means that the maximum transmit power of the non-high-gain mode have to be extended from 10 dBm to 23 dBm ($Ptx_{max,dBm} = 23$ dBm). Notice that $Ptx_{max,dBm}$ should be considered a parameter in this energy evaluation methodology and hence can be set to another value. The corresponding efficiency at $Ptx_{max,dBm}$ is scaled to be 40%, such that this model matches the assumption used in [3] for the constant efficiency model. This means that both models have the same power consumption at $Ptx_{max,dBm}$.

The resulting PA efficiency model proposed for MMTC devices, which has been derived from [12], is described in (4) and consists of two states; one where Ptx_{dBm} takes values from -30 dBm to 0 dBm where the power consumption is constant as in [12], and one where Ptx_{dBm} takes values from 0 dBm to 23 dBm where the power consumption increases at a moderate rate. Please note that (4), relates quantities given in dBm to the power consumption in W, similar to the model presented in [12].

$$P_{TX}(Ptx_{dBm}) = P_{PA}(Ptx_{dBm})$$
$$= \begin{cases} 0.0197 \cdot Ptx_{dBm} + 0.0454, & \text{if } 0 < Ptx_{dBm} \leq 23 \quad [\text{W}] \\ 0.0454, & \text{if } Ptx_{dBm} \leq 0 \end{cases} \quad (4)$$

The radio PA efficiency model for MMTC is depicted in Fig. 4 along with the commonly used constant PA efficiency model. The power consumption is given in relative values (in log scale) to the power consumption at $Ptx_{max,dBm}$. This new model will result in higher energy consumption if transmit powers lower than maximum transmit power ($Ptx \leq Ptx_{max}$) is used, as it models a lower efficiency than the constant model for $Ptx < Ptx_{max}$.

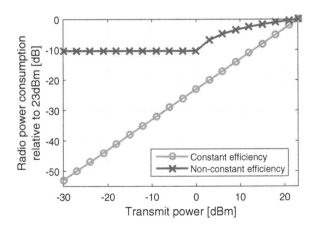

Fig. 4. Radio PA power consumption ($P_{PA}(Ptx_{dBm})$) for the proposed (non-constant efficiency) model for MMTC radios against the commonly used constant efficiency model (40%).

3   Results

This section demonstrates the use of the proposed energy evaluation methodology described in the previous section. Assumptions and parameters used throughout this section are listed in Table 1.

Table 1. Assumptions and parameters

Channel model	
Path loss (PL)	154 dB
Shadow fading	Eliminated by power control
Bandwidth	1.08 MHz
Noise	Noise figure (2 dB), thermal noise (-111 dBm)
Performance curve	
Target performance	10% error rate
γ_{SINR} (from Fig. 2)	Mode A (-18.9 dB), mode B (-11.0 dB)
Power model	
P_{TX}@23 dBm	500 mW
P_{other}	60 mW
T_{TX}	Mode A (0.903 ms), mode B (0.237 ms)
PSM	$P_{PSM} = 0.015$ mW, $T_{PSM} = 60$ min
$Idle$	$P_{Idle} = 3$ mW, $T_{Idle} = 2$ s
RX	$P_{RX} = 70$ mW, $T_{RX} = 0.36$ s
Power control	Uplink open loop with $\alpha = 1$
Traffic model	UL only. Poisson call inter-arrival
Deployment	Single cell. Path-loss compensated by PC
Access configuration	Transmission in all TTI. Consecutive repetitions

The assumptions intend to model a MMTC device which on average spends 60 min in PSM between consecutive uplink transmissions. Before the device is ready to transmit its payload it has to perform synchronization and read the needed control channel. Then it initiates the transmissions and stays in TX until all transmit repetitions have been performed. The values used in this evaluation are inspired by NB-IoT and should be considered as example values only. The power consumption values are from [7]. The time to conduct synchronization is from [5] and set to 200 ms which is spend in RX. The total acquisition time of the control channel is from [6] and is set to 2 s which is spend in $Idle$ followed by 160 ms in RX.

3.1   Impact of Power Control

To demonstrate the impact of uplink power control on the energy consumption, lets first have a look at Fig. 5 which shows the transmit power at different traffic

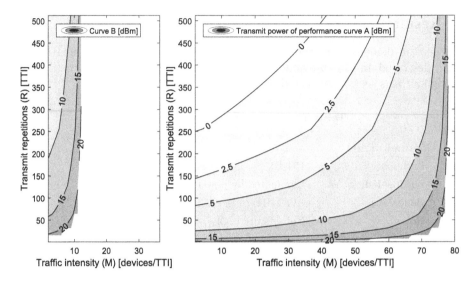

Fig. 5. Contour plot of the radio transmit power for performance curve A (right) and B (left). The figure for performance curve A can be interpreted as if 50 devices are transmitting at the same time ($M = 50$) and 64 transmit repetitions are configured ($R = 64$), then each device should transmit with $Ptx_{dBm} = 10\,dBm$ to reach the performance target (γ_{SINR}).

intensities (M) with the number of transmit repetitions (R) ranging from 1 to 512 for performance curve A (right) and B (left). Note that all devices are using the same number of transmit repetitions. The white area is the outage region and is clearly seen to the right and in the lower right corner for both performance curves. The outage region is where (2) is not satisfied meaning that the power control requests a transmit power above the maximum allowed (Ptx_{max}) and the UL γ_{SINR} cannot be met. Notice the outage region appears at much lower traffic intensity for performance curve B than for A. The reason for this is its higher γ_{SINR} which forces the transmit power to be generally higher than when performance curve A is used.

The corresponding energy consumption to the transmit powers shown in Fig. 5 is shown in Fig. 6. The markers in Fig. 6 at mode A (right) indicates the energy consumption if the transmit power is fixed to $Ptx_{dBm} = 10\,dBm$ meaning that power control is not utilized. Let's say that the traffic intensity at a given time is $M = 50$ devices/timeslot (TTI) and that $R = 64$ transmit repetitions is used. Then the traffic intensity increases to $M = 65$. To keep the target performance (γ_{SINR}) for the mode, in the case that power control is not utilized, meaning that the transmit power is fixed, the number of transmit repetitions have to be increased. The result is an increase in energy consumption. However, if power control is utilized, the transmit power can be increased with the result of maintaining the energy consumption instead and keeping the target performance. This is illustrated in the figure as the two arrows from $M = 50$

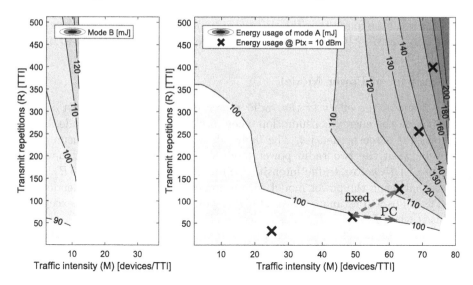

Fig. 6. Contour plots of the device energy consumption for mode A (right) and B (left) with power control as a function of transmit repetitions (R) and traffic intensity (M). The energy consumption is per transmission. The crosses in mode A are reference energy usage if the transmit power is fixed to 10 dBm. The arrows indicate the possible options when power control is used (blue) and not used (red) to keep the target performance (γ_{SINR}) when M increases from $M = 50$ with $R = 64$ to $M = 65$. (Color figure online)

to $M = 65$, one with power control (blue) and one with fixed transmit power (red). Note that it is the combination of the power model with the non-constant PA efficiency model for MMTC radios and power control that causes the non-linear energy consumption contour lines and enables new options to optimize the device energy consumption.

3.2 Impact of Performance Curves

Two modes are considered (A and B) each having a performance curve (A and B from Fig. 2) and transmit time. The target performance error rate is the same (10%), but the corresponding target SINR (γ_{SINR}) is different -18.9 dB for mode A and -11.0 dB for mode B (see also Table 1). The time it takes to transmit a single transmission is set to 0.903 ms and 0.237 ms.

The energy consumption of mode A and B are seen in Fig. 6. Again note the significant difference in maximum supported traffic intensity, which is much lower in B at $M \leq 12$ than in A at $M \leq 78$, as also seen in Fig. 5. The outage region appears much earlier with mode B due to the higher γ_{SINR} which is close to 10 dB higher. To compare the energy consumption of the two modes in further detail, one method is to extract energy consumption values at the same traffic intensities and transmit repetitions. If the transmit repetitions is fixed to

$R = 128$ it can be found that the energy consumption of B is slightly lower than A but only for $M \leq 10$ (e.g. at $M = 5$ the difference is 94.3 mJ against 97.4 mJ).

3.3 Impact of Power Model

Figure 7 shows the effect of the radio power model with a non-constant PA efficiency on the energy consumption when mode A is used. Using mode A and B shows the same tendencies. The figure shows that the energy consumption ratios between the two radio power models range from 1 to 2 and have an average of 1.15 across traffic intensity (M) and transmit repetitions (R). This means that using the power model with a non-constant PA model intended for MMTC radios will estimate a overall higher power consumption. The exception to this is when $Ptx = Ptx_{max}$, as expected. This is where the ratio is 1.

It should be noted that the energy consumption difference between the PA efficiency models only comes from the difference in energy consumption in TX state. But the energy consumption ratio depends on the energy consumption of the other states such as RX and $Idle$. For instance if the time spend in RX is increased to $T_{RX} = 1\,$s the average energy consumption ratio becomes 1.1 and the maximum ratio is 1.75. If T_{RX} is further increased to $T_{RX} = 2\,$s the average ratio is 1.07 and the maximum ratio becomes 1.5.

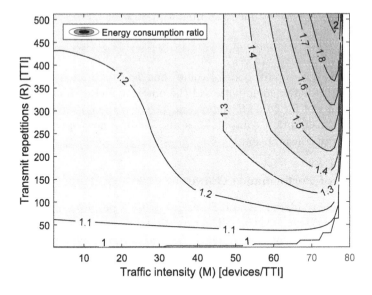

Fig. 7. Energy consumption ratio per transmission of the radio power model with the non-constant PA efficiency intended for MMTC devices over the commonly used radio power model with a constant PA efficiency model. Mode A is used.

4   Discussion

The results shown in this paper demonstrate the generic energy evaluation methodology and emphasize the impact of uplink power control, link-level performance and the power model. This section will discuss the results and the implications of the methodology.

One of the assumptions is the path-loss which is set to 154 dB in the results. The reason for selecting the path-loss as one value is that shadow fading is omitted as it is assumed that the power control is capable of compensating for this. Path-loss selected as a single value can also be interpreted as an upper bound of the shadow fading. Introducing fading as a random variable corresponds to introducing imperfect power control.

Perfect power control means, in this paper, that the devices are capable of doing perfect path-loss estimation and always have perfect knowledge of M and the path-loss. However in reality the power control might be inaccurate [10], in terms of e.g. path-loss estimation or knowledge of M. The effect of an imperfect power control will be that devices will select non-optimal transmit powers. Even if the average transmit power is the same as the optimal transmit power, the average amount of devices which fulfills the target performance will decrease. The overall consequence of a non-optimal transmit power is a lower outage capacity and a higher energy consumption.

The effect of using the non-constant PA efficiency model depends on the energy consumption of the transmit state compared to the other states (RX, $Idle$ and PSM) and the relative difference to the constant PA efficiency model. The relative difference of P_{TX} decreases as the transmit power increases (Fig. 4). However, when the transmit power increases the impact of P_{TX} in E_{tot} (3) increases. So when the transmit power increases on average, the average ratio between the power models (Fig. 7) will also increase as long that $Ptx < Ptx_{max}$. The maximum ratio will however not change if $Ptx = Ptx_{max}$ is already present.

Throughout the paper it is assumed that all devices uses the same number of transmit repetitions. Translated into a cellular deployment it will correspond to a group of devices which uses the same configuration, but are orthogonal to other groups and devices in the cell. If the path-loss is not the same for all devices in the group, the power control really proves its worth as it allows the devices to regulate themselves such that the received signal strengths after transmit repetitions at the BS receiver from the devices are still equally strong. When all devices in the group use the same number of transmit repetitions, the resource usage will also be fixed. This can, however, be optimized if devices are configured depending on their coverage conditions.

The proposed methodology is device centric and focuses on the uplink transmissions. The example evaluation done in this paper is for one uplink transmission being transmitted with transmit repetitions. The example considers what happens when the device is sleeping, has synchronized, read the broadcast information and control channel, received downlink traffic, transmitted its uplink transmission and returned back to sleep. It is possible to use this methodology for a protocol consisting of multiple uplink and downlink transmissions.

Downlink transmissions and their energy consumption impact are included as a parameter in the power model, where the most important parameter to change is how long time the device needs to be active. If the uplink transmissions utilize different modulation and coding scheme, multiple link-level performance curves will be needed.

The outcome of the proposed generic energy evaluation methodology is to make it easier to compare and evaluate standards and protocols for MMTC. The proposed methodology is simple, yet it includes important factors that affects the energy consumption. This means that evaluations performed with this methodology are more realistic than those done with the existing energy evaluation methodology from [3]. Further the outcomes of this paper can and should be used as input when new MMTC protocols and standards are being developed.

5   Conclusion and Outlook

This paper has presented a generic energy evaluation methodology tailored for MMTC. The methodology is demonstrated with special focus on three important factors which affect the energy consumption evaluation; power control, link-level performance and radio power model. The results presented in this paper provide important take-away messages:

- Using the commonly used radio power model with a constant PA energy efficiency instead of the radio power model with a non-constant PA model intended for MMTC radios can result in an overestimation of the battery life up to 100% and on average 15% across traffic intensity and transmit repetitions configurations. The proposed PA efficiency model does, however, need to be validated using a similar approach as used in [12] when NB-IoT or LTE-M devices become available.
- The combination of link-level performance, power control and the radio power model with a PA model intended for MMTC devices, provides options to optimize both access capacity and energy consumption.

Future work involves applying the proposed methodology on concrete protocols and help the development of future cellular MMTC solutions. These could be for example, new schemes and protocols such as one-stage and two-stage access protocols by [15]. For simplicity, in the presented evaluation, all devices are assumed to use the same radio access configuration. This can be generalized and interpreted as a group of devices within a larger set of MMTC devices. Our future work will focus on how cell radio resource management and higher layer protocol mechanisms can help minimizing the device energy consumption when several groups of devices are considered.

Acknowledgments. This work has been partly performed in the framework of the Horizon 2020 project FANTASTIC-5G (ICT-671660) receiving funds from the European Union. The authors would like to acknowledge the contributions of their colleagues in the project, although the views expressed in this contribution are those of the authors and do not necessarily represent the project.

References

1. 3GPP: Study on Enhancements to Machine Type Communications and Other Mobile Data Applications. TR 37.869 V12.0.0, September 2013
2. 3GPP: Study on Enhancements to Machine Type Communications and Other Mobile Data Applications Communications Enhancements. TR 23.887 V12.0.0, December 2013
3. 3GPP: Cellular System Support for Ultra Low Complexity and Low Throughput Internet of Things. TR 45.820 V2.1.0, August 2015
4. 3GPP: E-UTRA: Physical Channels and Modulation. TS 36.211 V12.8.0, December 2015
5. 3GPP: Narrowband LTE - Synchronization Channel Design and Performance. R1-156009, October 2015
6. 3GPP: NB-PBCH Design for NB-IoT. R1-160441, February 2016
7. 3GPP: UE Battery Life Evaluation for mMTC Use Cases. R1-165008, May 2016
8. Biral, A., Centenaro, M., Zanella, A., Vangelista, L., Zorzi, M.: The challenges of M2M massive access in wireless cellular networks. Digit. Commun. Netw. **1**(1), 1–19 (2015)
9. Fantastic-5G: Technical Results for Service Specific MultiNode/Multi-Antenna Solutions. Public Deliverable D4.1, H2020-ICT-2014-2, June 2016
10. Holma, H., Toskala, A.: WCDMA for UMTS: HSPA Evolution and LTE. Wiley, Hoboken (2007)
11. Jover, R., Murynets, I.: Connection-less communication of IoT devices over LTE mobile networks. In: 2015 12th Annual IEEE International Conference on Sensing, Communication, and Networking (SECON), pp. 247–255, June 2015
12. Lauridsen, M., Noël, L., Sørensen, T., Mogensen, P.: An empirical LTE smartphone power model with a view to energy efficiency evolution. Intel Technol. J. **18**(1), 172–193 (2014)
13. Laya, A., Alonso, L., Alonso-Zarate, J.: Is the random access channel of LTE and LTE-A suitable for M2M communications? A survey of alternatives. IEEE Commun. Surv. Tutor. **16**, 4–16 (2014)
14. Madueno, G., Stefanovic, S., Popovski, P.: Efficient LTE access with collision resolution for massive M2M communications. In: Globecom Workshops (GC Wkshps), pp. 1433–1438, December 2014
15. Saur, S., Weber, A., Schreiber, G.: Radio access protocols and preamble design for machine type communications in 5G. In: IEEE 49th Asilomar Conference on Signals, Systems, and Computers, November 2015
16. Zanella, A., Zorzi, M., dos Santos, A., Popovski, P., Pratas, N., Stefanovic, C., Dekorsy, A., Bockelmann, C., Busropan, B., Norp, T.: M2M massive wireless access: challenges, research issues, and ways forward. In: 2013 IEEE Globecom Workshops (GC Workshops), pp. 151–156, December 2013

Performance Evaluation of LAA-LTE and WiFi Coexistence in Unlicensed 5 GHz Band Under Asymmetric Network Deployments Using NS3

Maddalena Nurchis$^{(\boxtimes)}$ and Boris Bellalta

Department of Information and Communication Technologies,
Universitat Pompeu Fabra, 08018 Barcelona, Spain
maddalena.nurchis@upf.edu

Abstract. The tremendous growth of traffic demand over wireless networks is boosting a great deal of effort towards novel mechanisms for increasing the network capacity, in particular in cellular networks. One of the approaches currently under investigation is the carrier aggregation of LTE in licensed and unlicensed spectrum, which led 3GPP to include in the last Release the specification of Licensed Assisted Access (LAA) of LTE in the 5 GHz unlicensed spectrum. However, for LTE to fairly coexist with WiFi, many open issues still require investigation, as there is not yet a clear consensus on the actual impact that each network can have on the performance of the other.

In this paper, we evaluate the performance of LAA-LTE and WiFi when coexisting in the unlicensed spectrum under saturated traffic conditions and different network deployments. Our main contribution is the study of scenarios in which the two networks have unbalanced and asymmetric location, which allows to highlight how different AP/eNB location and mutual distance play a considerable role in the overall and individual performance.

Keywords: LTE · LAA · LBT · Coexistence · Evaluation

1 Introduction

Previous works have already demonstrated the need for a coexistence scheme of LTE with WiFi in the same spectrum, as it has been shown that the WiFi activity can be reduced to as low as 3% of the total time when LTE accesses the spectrum without implying any mechanism for fair sharing with other technologies already operating in the same spectrum [1–3]. In addition, regulatory requirements in Europe and Japan among others, do not allow LTE to access the spectrum without sensing the channel before. Many results from existing works have already highlighted that fairness of LTE when coexisting with WiFi should be obtained by designing LTE spectrum access schemes inspired by the WiFi paradigm. For instance, it has been shown that when LTE uses a Contention Window that does not increase exponentially, the impact on WiFi performance is noticeably negative compared to the case in which it does [2].

© Springer International Publishing AG 2016
T.K. Madsen et al. (Eds.): MACOM 2016, LNCS 10121, pp. 86–97, 2016.
DOI: 10.1007/978-3-319-51376-8_7

Despite the standardization process taken by 3GPP, which already led to the specification of LAA-LTE for downlink traffic, considerable work is needed for deeply understanding the interaction between the two technologies when both transmit in the same spectrum and the relation between benefits and limitations. Most of the existing works evaluate scenarios and traffic models mainly inspired by those specified by 3GPP in the preliminary feasibility study for LAA [4].

In this work, we analyse the performance of both WiFi and LAA-LTE both as standalone networks and when transmitting in the same spectrum. We focus on simple scenarios (e.g., with only one WiFi station (STA) or LTE User Equipment (UE) for each Access Point (AP) and eNodeB (eNB) respectively). We define our scenarios based on those specified by 3GPP, but modified in such a way to provide insights on cases that have not been deeply investigated so far. To the best of our knowledge, this is the first work that consider network deployments in which WiFi and LTE have different number of cells (i.e. AP or eNB) and that evaluate the impact of the inter-cell distance on the performance.

The rest of the paper is organized as follows. Section 2 describes the main principles of LAA-LTE and in particular of its channel access technique. We describe the LTE-WiFi coexistence implementation that we used for our evaluation in Sect. 3, while in Sect. 4 we specify the scenarios and the most important settings related to WiFi and LTE. Section 5 reports and analyses the results we obtained, and finally Sect. 6 concludes the analysis with some ideas for future work.

2 Licensed Assisted Access LTE (LAA-LTE)

The extension of LTE carrier aggregation protocol in unlicensed spectrum can be performed in several ways, some of them currently under discussion by the 3GPP consortium. LTE-U provides this extension without any carrier sensing mechanism. For this reason, this approach can not be implemented worldwide, due to the regulation of several markets, like Europe and Japan, where channel sensing is mandatory in unlicensed spectrum. Thus, this approach it is not being standardized by 3GPP, but technical specifications have been released by the LTE-U Forum. On the other hand, Licensed Assisted Access LTE (LAA-LTE), relies on the *Listen Before Talk* technique, which requires sensing the channel before accessing it for transmission. After approximately one year of study and discussion, 3GPP specified LAA for downlink operation in Release 13 and is currently working on additional features and uplink operation for Release 14. The current specifications have required a great deal of effort on the evaluation of the most efficient coexistence method [4]. Even within the LBT approach, at least 4 categories have been discussed, which can be summarized as follows:

- **LBT category 1:** no LBT
- **LBT category 2:** LBT with deterministic fixed channel sensing time
- **LBT category 3:** LBT with random backoff within a fixed size Contention Window (CW)
- **LBT category 4:** LBT with random backoff within a variable size CW

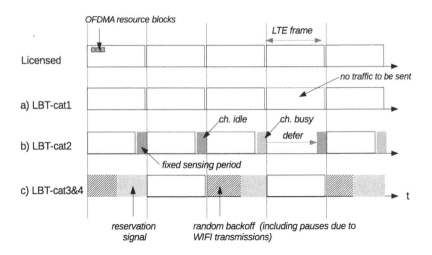

Fig. 1. LBT channel access scheme

Figure 1 shows each category in more detail and highlight how the channel is shared between LTE and WiFi. In the first case, LTE has no restrictions in accessing the channel whenever there are frames to be sent. On the other hand, LBT-cat2 may insert blank frames that can be exploited by WiFi to increase transmission opportunities. In fact, some existing works analyse the impact of the number of these frames on the performance of both networks, and propose to choose a pattern based on WiFi traffic load [5,6]. However, LBT-cat3 and LBT-cat4, which imply channel sensing before transmitting, have shown to be more suitable for reducing the difference in time the two technologies access the shared spectrum.

Existing works evaluating coexistence performance, have shown that the approach taken by LTE to access the channel should be the closest possible to the WiFi Distributed Coordination Function (DCF) mechanism, as the evaluation of other techniques that do not rely on this approach have shown the tendency to be unfair and provide advantage to LTE in accessing the spectrum [1,2]. For this reason, 3GPP standardization of LAA-LTE in Release 13 specifies an LBT scheme that closely resembles the carrier sensing and collision avoidance mechanism adopted by WiFi. Following these principles, in our evaluation we use an LBTcat4-based LAA-LTE as described in the following.

LAA-LTE relies on Clear Channel Assessment procedure (CCA) and Extended CCA (ECCA), with LAA Energy Detection (ED) threshold set by default to -72 dBm. When an eNodeB has packets to send, it first senses the channel for a CCA defer period of $43 \mu s$. If the medium is busy, it waits until it becomes idle again for the CCA duration period. When it is idle, it pick a random number within the current CW, starting with a CW of 15 (CWmin$_{LTE}$), and decrement the counter while the channel is idle, analogously to the WiFi DCF mechanism. Once the backoff counter expires, it can access the channel for transmission. In fact, LAA-LTE may rely on synchronization with the licensed LTE carrier, as it is not a standalone LTE over unlicensed. Hence, LTE can use a reservation signal if the channel is idle in order to keep it reserved from the time instant when the access is gained through LBT until the data subframe starting time is reached. We consider this approach, and we use one of the CW update rules specified by 3GPP [7], which is based on the number of NACKs received. More precisely, the CW is updated whenever the percentage of HARQ feedbacks reporting a NACK in the last transmission is at least 80%.

Finally, it is important to take into consideration that the most common way to define fairness in the context of LTE-WiFi coexistence, is the capability of LTE to cause a performance degradation that is not higher than the one that would be experienced in presence of another WiFi [4]. Although we believe that other metrics could be considered, for comparison with existing works we consider fairness based on this perspective during our evaluation, and leave the definition of alternative metrics as future work.

3   NS3 Coexistence Implementation

The increasing interest in extending LTE transmission also in unlicensed bands, and the fundamental need of evaluating the coexistence with WiFi, has boosted the development of custom simulators as well as additional modules to be included in existing tools, as in the case of NS3 [8]. NS3 is currently one of the most powerful and complete network simulator, open-source and with a strong community support. Motivated by the importance of the evaluation of WiFi and LTE coexistence in the same spectrum, the WiFi Alliance started a project for extending the simulator in order to include all the necessary elements to reproduce scenarios in which LTE and WiFi transmit in the same spectrum and are able to hear each other.

The main contribution is the implementation of the *LAA WiFi Coexistence module* [9], which implements new modules also according to the description of the coexistence evaluation scenarios specified by the 3GPP in [4]. In a nutshell, the coexistence module implements the elements necessary for the LTE physical layer to be connected to the same channel where WiFi devices transmit (in the 5 GHz band), and, at the same time, for the WiFi physical layer to detect and recognize *foreign* signals when sensing the channel. Moreover, the LTE channel access manager is extended so as to implement an approach based on the LTE duty cycling capability as well as a scheme reproducing the LBT and exponential backoff mechanism following the 3GPP RAN 1 design [10].

We would like to highlight that the main contribution of the module is the connection among the shared 5 GHz spectrum and the respective physical layer, which by default in NS3 are not able to interact or clearly detect each other. In particular, the *LBT Channel Access Manager* performs the carrier sensing and exponential backoff before accessing the channel, and the *Spectrum WiFi Physical* adds the capability to detect non-WiFi signals. The propagation model is also part of the coexistence module, and implements the IEEE 802.11ax indoor propagation model as part of the process of integrating NS3 with IEEE 802.11ax and other recent amendments. As for the WiFi standard, the default used in the coexistence module is IEEE 802.11n, but any other available in NS3 can be used.

4   Scenarios and Settings

Differently from other similar works, we do not strictly limit our evaluation on scenarios based on those defined by the coexistence evaluation analysis performed by 3GPP for LAA specification definition [4], but we define simple novel network deployments so as to get innovative insights on the performance and limitations of LTE(-LAA) and WiFi coexistence.

4.1   Scenarios

Reference Scenario. The first scenario we consider simply represents the basic case of one operator (e.g., WiFi or LTE) having only one cell, that is, one WiFi Access Point (AP) or one LTE eNodeB (eNB). We evaluate the performance when only one WiFi station (STA) or one LTE User Equipment (UE) is present. This case serves as a reference for each technology to obtain the maximum capacity achievable in the scenario of interest. This allows us to compare it with subsequent results obtained in scenarios where WiFi and LTE coexist in unlicensed spectrum under different deployment configurations. The top level of Fig. 2 illustrates the reference scenario.

Basic Coexistence Scenario. The first coexistence scenario includes only one cell for each network. The motivation behind this choice, is to rely on a simple case in which both technologies share the 5 GHz band, but limiting the parameters that may increase the uncertainty and the difficulty in understanding the reason behind the performance results obtained. Each cell consist of one AP/eNB, with a distance between them of d meters, as shown in the central part of Fig. 2, and all have the same Y coordinate. In our evaluation, we analyse how the performance of both networks change when varying this distance.

Asymmetric Coexistence Scenario. In order to have a more complete picture and more realistic results, we also consider the case of one operator having only one cell, and the other having two. With these settings, we aim to study the impact of each technology on the other under unbalanced conditions. To the best of our knowledge, this is the first work that employs this approach.

In order to better understand the interaction between the networks, we change the position of the cells and configure four different scenarios according to the cells configuration. In all the cases, as represented in the bottom part of Fig. 2, adjacent cells are placed d meters far from each other and all have the same Y coordinate. In order to compactly represent each case, the network each cell belongs to is represented by "W" for WiFi and "L" for LTE:

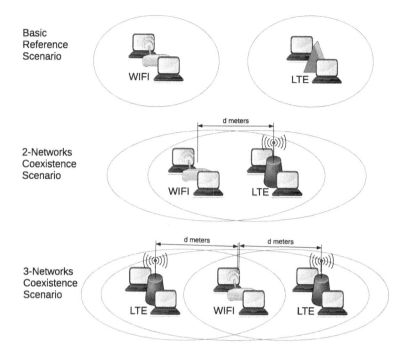

Fig. 2. Scenarios

- **WiFi-LTE-WiFi (WLW)** - LTE has only one cell placed in the center of the area, whereas WiFi has two cells placed at d meters of distance of the LTE cell at the two opposite sides.
- **LTE-WiFi-LTE (LWL)** - In this case, WiFi has only one cell placed in the center of the area, whereas the two cells of LTE are placed at d meters from the WiFi cell, analogously to the previous case.
- **WiFi-WiFi-LTE (WWL)** - The two WiFi cells are located next to each other and the only LTE cell is on the right side of the area, adjacent only the central WiFi cells.
- **LTE-LTE-WiFi (LLW)** - The two LTE cells are adjacent, and the one in the middle of the area is also adjacent to the WiFi cell, similarly to the previous scenario.

4.2   General Settings

The traffic model is Constant Bit Rate over UDP, with a rate of 200 Mbps. Based on previous simulations, we identify this rate as the one ensuring traffic saturation conditions. For each STA/UE associated to an AP/eNodeB, one traffic flow is setup for the whole duration of the simulation, which is set to 120 s. The source is outside the network and all the data traffic is downlink from AP/eNodeB to STA/UE. For each scenario, we run 5 independent replications and show the average of the obtained values.

Concerning the WiFi transmission rate and modulation, we set them fixed to **OFDM** and **24Mbps** respectively, so as to limit variability in the transmission conditions and focus on the performance variation. LAA-LTE performs carrier sensing before accessing the channel, through the LBT scheme described in Sect. 2.

Following the 3GPP guidelines for coexistence evaluation scenarios, in order to consider comparable settings, we set our area of interest as an indoor environment with size 120×50 m, hence we vary the distance d between adjacent cells from 10 to 110 m. Finally, in all scenarios, each STA/UE is located at a distance of 5 m from the AP/eNB, precisely having the same X coordinate and with Y coordinate 5 m smaller. To conclude this section, Table 1 shows the values of the main parameters related to LTE and WiFi.

Table 1. WiFi, LTE and general settings.

Parameter	Value
WiFi	
Transmission rate	24 Mbps
Transmission power	18 dBm
Modulation	OFDM
Bandwidth	20 MHz
Slot time	9 μs
ED threshold	−62 dBm
MAC protocol	EDCA
CW_{min}	15
CW_{max}	1023
LAA-LTE	
Channel Access	LBT w exponential backoff and reservation signal
Transmission power	18 dBm
CCA	43 μs
Txop	8 ms
Bandwidth	20 MHz
ED threshold	−72 dBm
CW update rule	NACKs 80% of HARQ feedbacks
CW_{min}	15
CW_{max}	1023
General	
Area size	120×50 m
Distance STA/AP and UE/eNB	5 m
Traffic model	CBR over UDP
Load	200 Mbps

5   Performance Evaluation

In this section, we present and discuss the results obtained in the scenarios previously described. In order to understand how the networks impact one on the other's performance, we limit our evaluation to the case of one user per cell, but we vary the distance d among adjacent cells from 10 to a maximum value of 110 m, so as to see the impact of the coexistence with decreasing distance in several scenarios, under symmetric and asymmetric network deployments. For each scenario, we report the normalized throughput of each network, which is calculated as the throughput achieved by each cell in a given scenario over the throughput in the reference scenario.

5.1   Basic Reference Scenario

This simple scenario is considered as a reference to know the maximum achievable capacity by each network under the conditions explained in the previous section. Table 2 reports the value of the throughput achieved by the node from the AP/eNB to the only STA/UE, showing a maximum throughput per user of 22 Mbps and 150 Mbps for WiFi and LTE respectively.

Table 2. Throughput of standalone networks

Scenario	Network	Throughput (Mbps)
Reference scenario	WiFi	22
Reference scenario	LAA-LTE	150

5.2   Basic Coexistence Scenario

In order to evaluate the coexistence in a simple case, similarly to the previous scenario, we consider only one cell for each network, namely, one AP and one eNodeB located in the same area of interest, and only one STA/UE per cell. In the following, we discuss the impact of the distance among neighboring cells on the performance of both networks.

Figure 3(a) shows the throughput of each cell in the case in which two WiFi networks coexist in the same area. Following the 3GPP approach, we evaluate this scenario in order to be able to compare the effect of LTE with the effect of another WiFi network under the same conditions. As mentioned in Sect. 2, this is in line with the common approach of evaluating the fairness as the capacity of LTE to impact WiFi not more than another WiFi. The results clearly show that a distance of 110 m is sufficient for the networks to not interact and thus reach the network capacity obtained in the case of a standalone network. On the other hand, while the distance between the two cells decreases, the aggregate throughput is nearly equally shared between the two cells.

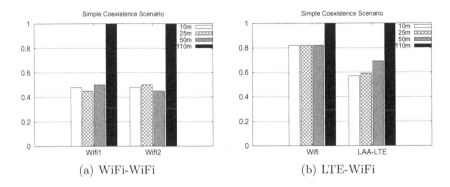

Fig. 3. Normalized throughput in the basic coexistence scenario

Figure 3(b) shows the performance of LTE and WiFi when they coexist under the same conditions of the WiFi-WiFi case. As other existing works, our results confirm that LTE can be a better neighbor than another WiFi in some scenarios. In this case, we can clearly see that for distances lower than 110 m, the LTE cell suffers a higher throughput degradation than the WiFi network. Indeed, the presence of a WiFi network in the vicinity has a stronger impact compared to the LTE standalone case, reducing the LTE throughput up to approximately 60% of the network capacity, in the case of 10 m distance. Also in this scenario, a distance of 110 m allows both networks to reach their maximum throughput as they do not interfere with each other.

5.3   Asymmetric Coexistence Scenario

To complete our analysis on the coexistence, we consider the case of different number of cells for each network, as described in Sect. 4.1. The main focus in the asymmetric deployments is the behaviour of the cell located in the middle, as it is expected to encounter more problems in accessing the channel, having always more contenders. Also in this case, we limit our analysis to the case of one active user per cell. In this section, we discuss the performance we observed in all the unbalanced scenarios and under different conditions in terms of distance among adjacent cells, in line with the previous section. In this case, being the length of the longest side of the area equal to 120 m, and having 3 cells to be placed within the area, we only consider the cases of d equal to 10, 25 and 55 m.

Two WiFi Cells - 1 LTE Cell. Starting with the case of two WiFi cells, Figs. 4(a) and (b) show the throughput of each cell in the case of the WiFi-LTE-WiFi and WiFi-WiFi-LTE configuration respectively.

In the **WiFi-LTE-WiFi** case, when the distance between adjacent cells is small, LTE can reach approximately half of its capacity. On the other hand, with higher distances, WiFi overall throughput increases and leads to a very poor LTE performance. Indeed, with a distance of 55 m, the two WiFi APs are far 110 m from each other, thus they do not contend with each other and contribute to starve the LTE cell, since it requires to detect both WiFi cells idle to initiate a transmission. On the other hand, when the cells are ordered differently as **WiFi-WiFi-LTE**, the two WiFi cells do interact with each other at all distances, and the impact on LTE performance degradation is limited to approximately half of its capacity. As expected, also in this case the cell that suffers highest throughput degradation is the cell in the middle.

Two LTE Cells - 1 WiFi Cell. With two LTE cells, the overall performance is significantly different than the case with two WiFi cells, as it is shown in Fig. 4(c) and (d). Regardless of the distance from LTE cells and of their location, WiFi never falls below 65% of its maximum throughput. Moreover, it reaches its maximum value when the WiFi cell is not in the middle and the distance with the LTE cell is such that they do not contend with each other for the channel. In this case, the LTE cell in the middle show a higher throughput loss than the others, similarly to other cases.

At this point, it is important to highlight that we set WiFi and LTE parameters to be inline with most of the existing works, thus the ED threshold of WiFi and LTE are different, as shown in Table 1. We can clearly see the effect of this in the last 2 figures, where WiFi is less affected than LTE under the same conditions (i.e. the throughput of the WiFi cell at the right edge of the

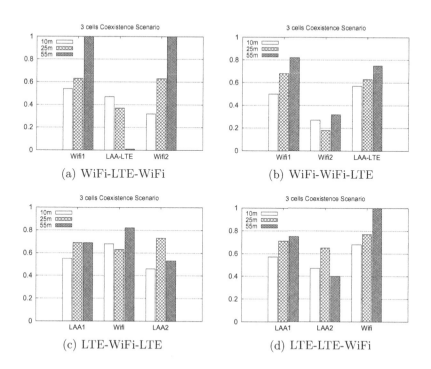

Fig. 4. Normalized throughput in the asymmetric coexistence scenario

LTE-LTE-WiFi scenarios reaches its capacity when the distance between cells is 55 m, whereas the throughput loss of the two LTE cells in the LTE-WiFi-LTE scenarios is higher.

From the results presented in this section, we can clearly observe that, by fixing the number of cells for each network and changing their relative position and distance, the variation of both networks' performance can be considerable. The main reason behind it is the interference between the cells and the effect of it on the channel access process. Surprisingly, the presented results show that WiFi is able to successfully survive when contending with even two LAA-LTE. In part, this is because WiFi is using a higher ED threshold, and therefore, it detects the channel empty more often, while LTE follows a more conservative approach.

6   Conclusion

In this paper, we present an evaluation of the performance of WiFi and LAA-LTE when they share the unlicensed 5 GHz spectrum under symmetrical and asymmetrical deployment settings. Our results confirm that in some scenarios LTE can impact WiFi performance less than another WiFi network. However, we also observe that the number of cells of each network, and their relative location and distance, have a considerable impact on the performance of both networks, as the interaction among them in accessing the channel changes accordingly. These findings confirm the importance of an adaptive channel access scheme able to guarantee fairness in the spectrum sharing based on the interference level between networks.

As a future work, we plan to investigate LTE-WiFi coexistence in more complex scenarios, with several spatially distributed networks and having some of them not in direct interaction with each other. Moreover, we are working on the definition of machine learning-based strategies to allow coordination among subsets of cells.

References

1. Sagari, S., Seskar, I., Raychaudhuri, D.: Modeling the coexistence of LTE and WiFi heterogeneous networks in dense deployment scenarios. In: IEEE International Conference on Communications (ICC), pp. 2301–2306. IEEE Press (2015)
2. Kwan, R., et al.: Fair co-existence of licensed assisted access LTE (LAA-LTE) and WiFi in unlicensed Spectrum. In: 7th Computer Science and Electronic Engineering Conference (CEEC), pp. 13–18. IEEE Press (2015)
3. Zhang, R., et al.: LTE-unlicensed: the future of spectrum aggregation for cellular networks. IEEE Wirel. Commun. **22**(3), 150–159 (2015)
4. 3GPP TR 36.889 V13.0.0 (2015–06). 3rd Generation Partnership Project; Technical Specification Group Radio Access Network; Study on Licensed- Assisted Access to Unlicensed Spectrum; (Release 13)
5. Almeida, E., et al.: Enabling LTE/WiFi coexistence by LTE blank subframe allocation. In: IEEE International Conference on Communications (ICC), pp. 5083–5088. IEEE Press (2013)

6. Zhang, H., et al.: Coexistence of Wi-Fi and heterogeneous small cell networks sharing unlicensed spectrum. IEEE Commun. Mag. **53**(3), 5083–5088 (2015)
7. 3GPP RP-151977 Status Report of WI Licensed-Assisted Access Using LTE (2015)
8. ns-3 web site. http://www.nsnam.org
9. ns-3 coexistence module. http://www.nsnam.org/wiki/LAA-WiFi-Coexistence
10. [RP-152233] (1, 2) 3GPP RP-152233 Status Report to TSG, Work Item on Licensed-Assisted Access to Unlicensed Spectrum (2015)

Providing Fast Discovery in D2D Communication with Full Duplex Technology

Marta Gatnau Sarret[1]([✉]), Gilberto Berardinelli[1], Nurul H. Mahmood[1],
Beatriz Soret[2], and Preben Mogensen[1,2]

[1] Department of Electronic Systems, Aalborg University, Aalborg, Denmark
mgs@es.aau.dk
[2] Nokia Bell Labs, Aalborg, Denmark

Abstract. In Direct Device-to-Device (D2D), the device awareness procedure known as the discovery phase is required prior to the exchange of data. This work considers autonomous devices where the infrastructure is not involved in the discovery procedure. Commonly, the transmission of the discovery message is done according to a fixed probability. However, this configuration may be not appropriate to meet the 10 ms control plane latency target defined for the next 5th generation (5G) system. In this work, we propose a distributed radio resource management framework supporting full duplex technology to provide D2D fast discovery. Such framework provides an algorithm to estimate the number of neighbor devices and to dynamically decide the transmission probability, for adapting to network changes and meeting the 10 ms target. Finally, a signaling scheme is proposed to reduce the network interference. Results show that our framework performs better than a static approach, reducing the time it takes to complete the discovery phase. In addition, supporting full duplex allows to further reduce the discovery time compared to half duplex transmission mode.

1 Introduction

Device-to-Device (D2D) communication has drawn significant attention for the design of 5th generation (5G) systems to offload the infrastructure and to cope with the continuous growth of wireless applications and services. In D2D communication, devices are allowed to communicate directly, without the involvement of the infrastructure. However, prior to the establishment of such communication, devices must discover their peers. This device awareness procedure is known as discovery phase. According to the latest specifications for next generation systems [1], the control plane latency cannot exceed 10 ms. Such requirement poses challenges on how to facilitate fast device discovery. Full duplex (FD) technology, which allows for simultaneous transmission and reception in the same frequency band, may speed-up the discovery process.

The execution of the D2D discovery procedure can be controlled by the infrastructure or performed autonomously by the devices. The first option requires the exchange of messages with the base station, generating additional

© Springer International Publishing AG 2016
T.K. Madsen et al. (Eds.): MACOM 2016, LNCS 10121, pp. 98–108, 2016.
DOI: 10.1007/978-3-319-51376-8_8

control overhead and increasing the latency. The latter option, where devices send the discovery message periodically, has potential to diminish the control overhead and provide lower latency [2].

Autonomous device discovery using conventional half duplex (HD) transmission mode has been studied by the research community [4, 5, 7], whereas few works consider FD technology [8–10]. A synchronous distributed ad-hoc network is studied in [5], focusing on optimizing the discovery latency and the number of discovered devices. The authors propose a resource structure as well as a resource selection. However, the feedback mechanism is not considered and their system operates in a larger time scale than that specified by [1]. A discovery message design to minimize collisions is proposed in [4]. The work analyzes an autonomous D2D system where devices transmit the discovery message with a fixed probability, showing an improvement in the number of discovered devices. The authors in [7] propose using a small portion of the resources for new devices appearing in the network, such that their discovery message can be transmitted with a shorter delay. In [8], a strategy to reduce idle slots and collisions is presented. FD is used to detect the activity of other devices. The work assumes that a device stops transmitting when it is discovered. Nevertheless, this assumption may not be valid in networks with dynamic (de)activation of the nodes where transmitting the discovery message is always required. In [10], FD is combined with compressed sensing to overcome the drawbacks from HD and single packet reception. The authors claim that the discovery phase is completed in a single time slot. However, a very limited number of neighbors is considered and the feedback procedure for discovery acknowledgment is not addressed. The authors in [9] evaluate FD with directional antennas, where each device selects a transmission direction randomly at each time slot. It is important to notice that the mentioned works assume the transmission of the discovery message with a fixed probability. This principle does not allow to control the generated idle slots and collisions, thus posing critical challenges in meeting the latency requirements.

In our previous work [2], we showed that adapting the rate of discovery message transmission to the number of active devices may be beneficial, and we identified challenges in terms of interference management in a large network. In this paper, we propose a radio resource management (RRM) framework for autonomous D2D communication supporting FD technology. It provides a mechanism to estimate the number of neighbors as this information is not available in realistic ad-hoc networks, and an adaptive scheme to select the most appropriate transmission probability for the discovery messages. The interference can be better coordinated by allowing the devices to exchange their transmission probability, which captures the number of neighbors in our proposal. Thus, each terminal can dynamically set the most appropriate transmission probability using not only the current value and own information but also information from the neighbors. Results show that our solution achieves lower latency than a static approach. Moreover, supporting FD allows to further reduce the discovery time compared to HD transmission mode.

The paper is organized as follows. Section 2 describes the proposed RRM framework. Section 3 presents the system model and discusses the simulation results. Finally, Sect. 4 concludes the paper and states the future work.

2   D2D Fast Discovery

2.1   General System Overview

We focus on autonomous ad hoc networks with a dedicated band of the spectrum for the discovery procedure. Devices communicate directly with each other and the infrastructure is not involved in the discovery phase, but still provides time and frequency synchronization. This design allows to avoid interference between cellular and D2D users.

A time slotted system is considered. At each transmission opportunity, there is possibility of exploiting a pool of orthogonal frequency resources, where the resource to be used is randomly chosen. It is assumed that, on reception, devices can simultaneously listen to all frequency resources. The discovery message is transmitted in a broadcast manner according to a certain transmission probability ρ and it contains the information required to perform the discovery phase, e.g., the device identifier and its position. Since the discovery procedure needs to be completed in a short time to meet the strict control plane latency requirements [1], the number of link failures should be minimized. This can be achieved by transmitting the discovery message with a robust modulation and coding scheme (MCS) at the expense of a larger message, and by using one spatial stream, often referred as transmission *rank* one, assuming that devices are equipped with 4×4 multiple-input multiple-output (MIMO) transceivers. The dimensioning of the discovery message is left for future work. In this paper, we assume that the discovery message can be mapped over a single time/frequency resource.

The discovery phase is required to set a unicast/multicast communication. Therefore, the devices involved in such communication should be acknowledged of the fact that their peers are aware of their presence. We propose a design for the discovery message that includes a *feedback* field, containing the identifiers of the devices that have been discovered by the transmitting device. Since the discovery message is broadcast, a device that receives and decodes the message will check if its identifier is piggybacked. If so, the receiving device will know that it has been discovered by the transmitting device. The discovery time is then based on the feedback reception time, and it depends on the transmission probability ρ. Using a high ρ causes a large number of collisions which increases the discovery time. On the other hand, using a small ρ creates a large number of idle slots due to the inactivity of the devices, which also increases the time needed to complete the discovery procedure. Furthermore, in case of HD transmission, the necessity of transmitting the discovery message leads to a reduction of the opportunities for listening to neighbors' transmissions. We investigate the potential of FD technology in reducing the discovery time, since it eliminates the HD constraint by allowing simultaneous transmission and reception on the same frequency band.

2.2   RRM Design

In our previous work [2], we showed that the transmission probability that leads to the minimum discovery time depends on the scenario, e.g., on the number of neighbors. Such result indicates that a dynamic choice of ρ can be beneficial for the system. In addition, we identified challenges in terms of interference management in large networks. Let us define *cluster* as the set of neighbors within the coverage range of a device, plus the own device. Therefore, the cluster and its size is a device-specific parameter. In case every device is able to reach all the other devices in the network, all the devices' clusters coincide. We refer to this case as *single cluster* network. The opposite case is a *multi-cluster* network. Figure 4 shows an example of a portion of a multi-cluster network, where the clusters from two devices, C and G, are highlighted. In particular, the number beside each device refers to their cluster size. In this specific example, C only reaches G, while the latter reaches C, Y and W. Let us focus on G, which has two neighbors perceiving a larger cluster size (W and Y) and C, which only reaches G. Since G is not aware of the overall interference perceived by W and Y and it has only three neighbors, it would benefit from using a high ρ. However, using a high ρ may increase the number of collisions to W and Y, who have a larger number of neighbors, consequently increasing their discovery time. On the other hand, C will benefit from such a high ρ because it has a cluster of size 2.

From the previous example we can extract that an exchange of information among devices can be beneficial to reduce the overall network interference and to avoid increasing the discovery time. Furthermore, in a realistic network, the information related to the number of neighbors is not available. To solve the mentioned problems, we propose a RRM framework to dynamically adjust the transmission probability allowing devices to adapt to network changes. It consists of two parts: the instantaneous estimation of the number of neighbors, and the dynamic adjustment of ρ based on network information exchange. The proposed solution is distributed, so it does not require a centralized controller that collects information from the network.

Pseudo Code 1. Algorithm for estimating the number of neighbors

---

$\rho \leftarrow$ Current transmission probability. Initial value $\rho = 0.5$
$\tilde{M} \leftarrow$ Estimated number of neighbors. Initial value $\tilde{M} = 0$
repeat   At each time slot
    Extract ρ according to the selected information exchange
    approach (Table 1)
    if Transmission time, based on ρ, **then**
       Transmit the discovery message
    else
       Receive discovery messages from neighbors, and
       estimate $\tilde{M}$ as:

$$\tilde{M} = \frac{\#\text{decoded signals}}{\rho} \tag{1}$$

    end if
until The device turns off

Estimating the Number of Neighbors. The estimation of the number of neighbors is done based on the available information at each device: the own ρ and the number of signals successfully decoded in each receiving time slot. The estimation of the number of neighbors with HD is done as described in Pseudocode 1. In case of FD, the *if* statement encapsulates only the transmission part, since a FD device is continuously receiving. In Eq. 1, the *#decoded signals* refers to the number of instantaneous messages that a device on reception can successfully decode. Then, the equation is equivalent to the number of active neighbor devices that a node can detect.

Signaling Scheme. To reduce the network interference, we propose that devices send ρ within the discovery message, since it is related to the number of estimated neighbors: a low number of estimated devices leads to a high transmission probability, and vice versa. The value of ρ can be represented with different number of bits, depending on the desired resolution and the allowed control overhead. Table 1 lists the proposed approaches to extract ρ, according to the signaled information. The difference among these approaches is the amount of extra information sent within the discovery message and how this information is utilized to decide the ρ to be used. With the *selfish* approach, devices do not signal any information about their ρ, and they behave in a selfish

Table 1. Information exchange approaches

Approach	Signaling	Principle
Selfish (ρ_{sf})	None	Use ρ extracted from the estimated number of devices without considering information from the neighbors. Section 3.1 describes the function that, depending on the estimated number of devices, provides the most appropriate ρ.
Cooperative minimum	ρ_u	Set ρ as the minimum between: a. The ρ extracted from the estimated number of neighbors (ρ_{sf}) b. The minimum ρ received from the neighbors
Cooperative maximum	ρ_u and ρ_{sf}	Set ρ taking into account: a. The ρ extracted from the estimated number of neighbors (ρ_{sf}) b. The used ρ received from the neighbors $(\rho_{u,nb})$ c. The estimated ρ received from neighbors $(\rho_{sf,nb})$ The decision is taken as follows: 1. Extract the minimum of $\rho_{u,nb}$ and $\rho_{sf,nb}$ of all the received messages from my neighbors 2. If $\rho_{u,nb} = \rho_{sf,nb} = \rho_{nb}$: select the minimum between ρ_{sf} and ρ_{nb} 3. If $\rho_{u,nb} < \rho_{sf,nb}$: select the minimum between $\rho_{sf,nb}$ and ρ_{sf}

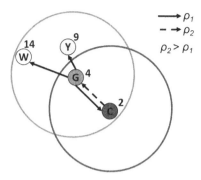

Fig. 1. Example of a multi-cluster network with 2 highlighted clusters (Color figure online)

manner. Hence, the control overhead is not increased but the network interference is uncontrollable. With the *cooperative minimum* option, devices signal the used ρ, which is extracted as indicated in Table 1. In this case, the control overhead is slightly higher but the interference is reduced. Finally, with the *cooperative maximum* approach, devices transmit two values of ρ: the one used for transmission, extracted by applying the *cooperative maximum* approach, and the one extracted from the estimation of the number of neighbors, i.e., the *selfish* ρ_{sf}. The difference between the second and the third approaches is that the latter avoids for the minimum ρ to spread across the network, at the cost of limited extra overhead.

Let us focus again in the example depicted in Fig. 1. Using the *selfish approach*, G and C will use a high transmission probability, hence G will be highly interfering W and Y, causing unsuccessful transmissions. If the *cooperative minimum approach* is used, the ρ extracted by W will be the one that the other devices in the network will use. However, C should be able to transmit with a higher ρ because it is not interfering W. This situation can be solved with the *cooperative maximum approach*.

3   Performance Evaluation

The RRM framework techniques proposed in Sect. 2 are evaluated using our own developed Matlab simulator. Firstly, we want to prove that a dynamic ρ selection has advantages over using a fixed one. Secondly, the performance of the single and the multi-cluster scenarios is discussed.

For our D2D network, we assume a simple path loss model $\gamma = d^{-\alpha}$, where α is set to 4, and no fading. The packet decoding is threshold-based: a collision occurs if the signal to noise-plus-interference ratio (SINR) is below such threshold, which is set to 0 dB, assuming Quadrature Phase Shift Keying (QPSK) with a coding rate of 1/3 and a block error rate of 0.01%. The representation of ρ is ideal in this study, i.e., it is represented with maximum resolution. Finally, ideal self-interference cancellation in FD devices is assumed [3].

According to the findings in [2], one of the requirements to meet the strict 5G control plane latency target is to use interference cancellation (IC) receivers. These receivers are able to suppress the $N - K$ strongest interfering streams, where N is the number of MIMO receive antennas and K is the transmission rank. For example, according to our simulation setup (transmission rank one and devices equipped with 4 receiving antennas), the three strongest interfering streams can be suppressed. We assume ideal IC in this work. The SINR reads then:

$$SINR_{IC} = \frac{\gamma_d \cdot P_T}{\sum_{i=k}^{rxSignals} I_i + N_0} \qquad (2)$$

where γ_d is the pathloss between the receiver and the desired transmitter, P_T is the transmit power, N_0 is the noise power, $k = N - K + 1$ and the interfering streams I_i are sorted based on the signal strength from the strongest to the weakest one. In this work, the transmit power is 0 dBm and a noise power in each time/frequency resource is -95 dBm.

The devices are randomly deployed in a certain area. Two scenarios are analyzed, a single and a multi-cluster scenario. The former suffers only from intra-cluster interference, while the latter is affected by intra and inter-cluster interference, since clusters may be partially overlapped or totally isolated. The single cluster scenario refers to a $100 \times 100 \, \text{m}^2$ area with the number of deployed devices ranging from 10 to 50. An area of $1000 \times 1000 \, \text{m}^2$ is considered for the multi-cluster scenario, where the number of deployed devices goes from 10 to 300. In particular, the average cluster size ranges from 3 to 44.

We assume that a discovery message opportunity occurs every 0.25 ms. This is consistent, for example, with the assumption in terms of frame duration of our 5G small cell concept presented in [6]. The discovery time is defined as the time needed for a device to be discovered by all its neighbors, based on the feedback reception time. It is extracted, individually for each node, as the maximum time among all the neighbor feedback reception times. For the single cluster scenario, results are presented in terms of average discovery time, since the interference conditions for all the devices are, in average, the same. In case of the multi-cluster scenario, results are presented in terms of the 95th percentile of the discovery time. For both scenarios, the performance of FD is compared against the HD performance. Finally, the variable θ provides an indication of the system congestion, and it is defined as:

$$\theta = \frac{\text{number of network devices}}{\text{number of frequency resources}} \qquad (3)$$

3.1 Dynamic Transmission Probability

Let us consider the single cluster scenario with its corresponding parametrization, and setting the size of the frequency resource pool to 1, 2 and 4, to have different system level congestion representations. In our previous work we proved that the optimal ρ which minimizes the discovery time is scenario-dependent. Figure 2 shows such optimal ρ as a function of θ, for the ideal case where the devices know the exact number of neighbors. From the figure we can observe that, as the system congestion increases, the optimal ρ diminishes due to the constraint on the number of collisions. We can also see that FD allows for a higher ρ in some cases, specially at low system congestion. This is because, as explained in Sect. 2, FD solves the constraint that HD poses on the ρ selection. Therefore, since FD operates in a larger range of ρ, we expect to have larger gains from using a dynamic approach with FD compared to HD. This two curves can be easily approximated by a simple multiplicative inverse function $\rho = f(\tilde{M}, \text{frequency resources pool size}) = f(\theta) \approx \frac{a}{\theta^b}$, where a and b are fitting parameters. Such approximations allow us as to have a representation of the optimal ρ based on the system congestion.

3.2   Single Cluster Performance

As the next step, we evaluate the proposed RRM framework in the single cluster scenario, assuming a frequency resource pool of size 4. In this case, the considered approach is the selfish one, since the interference conditions of all the devices are, in average, the same. Consequently, the three approaches described in Table 1 show the same performance. The evaluation is done by comparing: the optimal discovery time, extracted under the assumption of ideal information at the devices; the performance of the proposed RRM framework, extracting ρ_{SF} from the inverse approximation of the curves shown in Fig. 2; and a fixed ρ of 40%, since it provides a good trade-off on the HD performance given its limitation of not being able to receive messages while transmitting. Figure 3 shows the performance comparison between the mentioned cases. We can observe that the HD performance is barely affected from the usage of a dynamic ρ, except when the number of network devices is large. This is caused by the small operational range of ρ, given the HD constraint. However, in case of FD, we can observe that a dynamic ρ selection allows to get very close to the optimal system performance. The maximum difference between the optimal and the algorithm performance is 0.83 ms, at high congestion. At low congestion, the maximum difference is 0.13 ms.

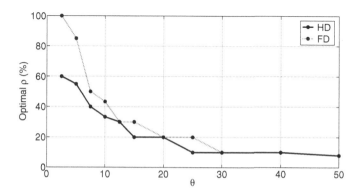

Fig. 2. Optimal ρ as a function of the system congestion θ

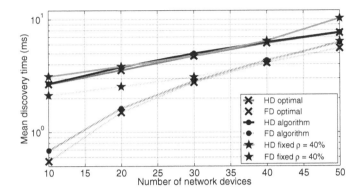

Fig. 3. Single cluster. Performance comparison between the proposed algorithm, the optimal performance and a fixed ρ

We want to emphasize that the robustness of the approximated ρ curves have also been evaluated, by varying the path loss model or the deployed area. Results show that the maximum difference between the ideal discovery time and the one extracted by using our proposed framework and the approximations is $\sim$0.5 ms.

3.3   Multi-cluster Performance

We focus here on the multi-cluster scenario, assuming a frequency resource pool of size 4, and we will analyze the performance of the three approaches presented in Table 1 and the fixed 40% transmission probability. For HD, only two performance curves are presented, the selfish and the fixed 40% one, since the performance with the other two proposed algorithms is nearly the same as the selfish one. The reason for that, as explained before, is the smaller range of ρ with HD since a device cannot listen while transmitting, leading to an increase of the discovery time if a too high ρ is used. Figure 4 shows the 95th percentile of the discovery time. The results show that, with HD, the latency requirement of 10 ms can be achieved with up to 200 network devices, but the discovery time is always larger than the one achieved with FD. Focusing on the FD performance, we can observe that providing a dynamic solution for the ρ selection brings benefits in terms of reduction of the discovery time, independently of the network density. In case of few devices in the network, the selfish approach shows the best performance in terms of discovery time, since it is the most aggressive scheme and using interference cancellation receivers allows the system to use a high ρ. The cooperative minimum approach is too conservative at low network density, since the ρ used by the device in the worst interference conditions spreads across the network. Note that the drawback of such approach can be solved by using the cooperative maximum approach, since it allows to increase the transmission probability in case the neighboring nodes are not affected by this ρ increase. As the network density increases, we observe that the cooperative minimum and maximum approaches reach the optimal system performance. In this case,

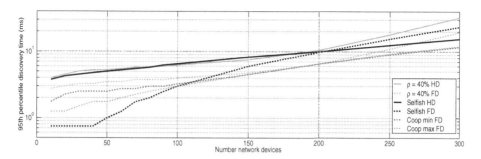

Fig. 4. Multi-cluster. Performance comparison between the proposed algorithm, the optimal performance and a fixed ρ

both approaches perform equally since the optimal ρ is already small, to avoid generating a large interference and hence a large number of collisions. Notice that with the proposed cooperative algorithms, the system can support higher network density while still meeting the 10 ms latency requirement.

Finally, Fig. 5 shows the cumulative distribution function (CDF) of the discovery time with HD and FD for the cooperative maximum approach, for the cases of 70 and 150 number of network devices. The curves show that, beyond the mean gain of FD over HD, HD suffers from larger time variances, with some devices perceiving large discovery time and others finishing their process much faster.

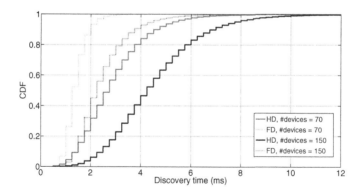

Fig. 5. Multi-cluster. Cooperative maximum algorithm CDF

4   Conclusion and Future Work

In this paper we have proposed a RRM framework supporting FD technology to provide fast discovery in autonomous D2D communication. The framework provides a mechanism to estimate the number of neighbor devices and a scheme to dynamically adjust the transmission probability, in order to reduce the number of collisions and idle slots and decrease the latency for discovery. The proposed signaling exchange mechanism allows to reduce the network interference and improve the system performance. System level results show that our proposed dynamic solution allows to reduce the discovery time compared to traditional static approaches, especially in case full duplex technology is used. Future work will focus on the control message design and on applying the framework in scenarios with high mobility such as vehicle-to-anything (V2X) applications.

Acknowledgments. This work has been performed within the framework of the Horizon 2020 project FANTASTIC-5G (ICT-671660) receiving funds from the European Union. The authors would like to acknowledge the contributions of their colleagues in the project, although the views expressed in this contribution are those of the authors and do not necessarily represent the project.

References

1. 3GPP TR 38.913 V0.3.0: Study on scenarios and requirements for next generation access technologies, March 2016
2. Gatnau, M., Berardinelli, G., Mahmood, N.H., Soret, B., Mogensen, P.: Can full duplex reduce the discovery time in D2D communication? In: 13th International Symposium on Wireless Communication Systems (ISWCS), Accepted June 2016
3. Heino, M., et al.: Recent advances in antenna design and interference cancellation algorithms for in-band full duplex relays. IEEE Commun. Mag. **53**(5), 91–101 (2015)
4. Hong, J., Park, S., Choi, S.: Neighbor device-assisted beacon collision detection scheme for D2D discovery. In: 2014 International Conference on Information and Communication Technology Convergence (ICTC), October 2014
5. Jung, S., Chang, S.: A discovery scheme for device-to-device communications in synchronous distributed networks. In: 16th International Conference on Advanced Communication Technology, February 2014
6. Mogensen, P., et al.: Centimeter-wave concept for 5G ultra-dense small cells. In: IEEE 79th Vehicular Technology Conference (VTC) (2014)
7. Park, S., Choi, S.: Expediting D2D discovery by using temporary discovery resource. In: IEEE Global Communications Conference (GLOBECOM), December 2014
8. Sun, G., Wu, F., Gao, X., Chen, G.: PHED: pre-handshaking neighbor discovery protocols in full duplex wireless ad hoc networks. In: IEEE Global Communications Conference (GLOBECOM), December 2012
9. Xu, R., Li, J., Peng, L., Ye, Y.: A neighbor discovery algorithm for full duplex ad hoc networks with directional antennas. In: The 27th Chinese Control and Decision Conference (CCDC), May 2015
10. Yang, X., Wang, X., Yang, R., Zhang, J.: Full-duplex and compressed sensing based neighbor discovery for wireless ad-hoc network. In: IEEE Wireless Communications and Networking Conference (WCNC) (2015)

Information Theory

A Secrecy and Security Dilemma in OFDM Communications

Andrey Garnaev[1,2(✉)] and Wade Trappe[2]

[1] Saint Petersburg State University, St. Petersburg, Russia
garnaev@yahoo.com
[2] WINLAB, Rutgers University, North Brunswick, USA
trappe@winlab.rutgers.edu

Abstract. Emerging wireless systems will need to address multiple, conflicting objectives: ensuring the secrecy of the communication secrecy (i.e., maximizing the un-eavesdropped throughput at the receiver), and ensuring the "security" of the communication (i.e., minimizing damage that might arise because of the possibility that there is leakage in communication). This problem presents a dilemma for system engineers since the optimal solution for one of objective might be not optimal for the other. This dilemma calls for a need for designing a trade-off solution for these objectives. In this paper, we propose the use of a Kalai-Smorodinsky bargaining solution between secrecy and security for OFDM-style communications, which allows one to select the security level of communication. In particular, this solution allows one to maintain the ratios of maximal gains for cooperative fulfilment of both objectives.

Keywords: Bargaining · Eavesdropping · Information leakage

1 Introduction

The problem of establishing secure communication is a fundamental challenge in wireless systems, and it involves multiple aspects that need to be addressed simultaneously. One challenge is to ensure the *secrecy* (or confidentiality) associated with the communication information, i.e., appropriately encoding communication so that there is a guarantee regarding the rate at which data can be secretly conveyed in the presence of an eavesdropper. A different aspect, which we call the *security* of the communication session, deals with minimizing the risk or damage associated with the possibility that an adversary observes some of the communication, and hence a second design objective is to minimize the leakage (regardless of whether there was coding involved) or that the communication is detected by the adversary.

These problems have received significant attention in the literature over the past decade. A comprehensive survey of secrecy and security in multiuser wireless networks is given in [24]. As examples of secrecy, we mention just a few of

© Springer International Publishing AG 2016
T.K. Madsen et al. (Eds.): MACOM 2016, LNCS 10121, pp. 111–125, 2016.
DOI: 10.1007/978-3-319-51376-8_9

works. For models of an interference channel with an external eavesdropper [19], for secret communications over fading channels [22] and over a fading eavesdropper channel [20]), for employing artificial noise to improve secret communication [25], for an adversary combining eavesdropping and jamming attack in CDMA (Code Division Multiple Access) networks [10,31] and in OFDM (Orthogonal Frequency-Division Multiplexing) (for the low signal-to-interference-plus-noise ratio regime) networks for one-time slot attack [8] and dynamic attack [11], for spectrum coexistence in dynamic spectrum access [12], for improved physical layer security in amplify-and-forward (AF) relay networks with two-hop information leakage [23], for unknown eavesdropping capability of the adversary [9], on the benefits of using directional antennas in an eavesdropping attack [18], and the silent mode against a combined eavesdropping/jamming attack [7], for unknown eavesdropping purpose [6], for improving secrecy with the help of trustworthy secondary users in cognitive radio networks [27], for detecting pilot spoofing attack in multiple-antenna systems [28].

As examples of security, we mention just a few of works. Effect of correlated sources across an eavesdropped channel that incorporate a heterogeneous encoding scheme on the information leakage when some channel information and a source have been wiretapped is investigated in [3]. To estimate leakage in a rich variety of operational scenarios a generalization of min-entropy leakage was introduced in [2]. A secrecy rate enhancing relay strategy supporting information leakage neutralization for the multi-antenna non-regenerative relay-assisted multi-carrier interference channel was proposed [16]. Bandwidth monitoring with uncertainty on the type of adversary's activity was modeled [13]. A cooperative power allocation algorithm to mitigate security vulnerability by limiting the signal to interference and noise ratio at the eavesdropper receiver was suggested in [4]. In [14], a tiling based scanning algorithm for detecting an intruder signal in a wide amount of bandwidth was proposed.

In this paper, *first* we propose to consider the problem of designing a secure communication as two objectives (namely, secrecy and security) problem. As the objective for secrecy communication we use the secrecy rate, while the probable leakage in the communication is used as the objective for communication security. The challenge that is inherent in such two objectives problem is that the optimal solution for one objective might be not optimal for the other. *Second,* to solve this dilemma, we generalize a classical problem of secret communication by introducing a notion of security level for communication. *Third,* we explicitly solve this generalization in closed form for a fixed security level. *Finally,* we find the optimal security level as a Kalai-Smorodinsky bargaining solution, which maintains the ratios of maximal gains for cooperative fulfilment of the both objectives. A survey of different bargaining concepts used in wireless communication is given in [29].

The organization of this paper is as follows: in Sects. 2 and 3, we first give auxiliary results associated with the communication in secrecy and security modes. In Sect. 4, the notion of security level is introduced and the optimal transmission protocol for a fixed security level is designed. In Sect. 5 an iterative algorithm for finding the optimal strategy is given. In Sect. 6, we extend our work to the problem of bargaining over the security level, and find a solution for the secrecy versus security bargaining problem. Finally, in Sect. 7, and the Appendix, a discussion of the results as well as the proofs of the results are offered.

2   Secrecy Mode

In this section, we briefly consider the basic secrecy communication model, which involves a protagonist, Alice, who wishes to communicate secretly with her colleague, Bob. At the same time, Eve is a passive adversary eavesdropping on the communication between Alice and Bob.

This basic secret communication scenario is portrayed in Fig. 1. In this work, the underlying wireless medium has been channelized into n separate channels (e.g. different subcarriers in an OFDM system). Thus, Alice communicates to Bob across n channels, and the channel responses for these n channels are represented by coefficients h_i, $i \in \{1, \dots, n\}$. We assume (as in a wireless setting) that this communication occurs across a broadcast medium, and that Eve can eavesdrop on Alice's communication. The channels from Alice to Eve are represented by coefficients h_{Ei}. The signal transmitted by Alice as X, the signal received by Bob is Y and by Eve is Z.

A complete characterization of the secrecy capacity was provided in [5], and is given by

$$CS = \max_{V \to X \to YZ} I(V;Y) - I(V;Z),$$

where V is an auxiliary input and $I()$ is the mutual information. The secrecy rate $C = \max_X I(X,Y) - I(X,Z)$ is often studied in preference to the secrecy capacity since there is currently no systematic approach to optimize the secrecy capacity over the auxiliary input V. Nevertheless, when Bob's channel is more capable than Eve's channel, the secrecy capacity and secrecy rate are the same. In this paper, we relax assumption that all the channels have to be eavesdropped. Namely, we assume that a channel i could be eavesdropped with probability q_i where $q_i \in [0, 1]$, and Alice knows these probabilities, as well as all the channel's gains. The notion of q_i is introduced to reflect the fact that Eve might have

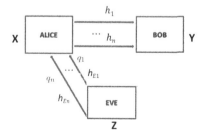

Fig. 1. Relationships between Alice, Bob and Eve in secrecy communication.

limited resources to devote to the attack, or might have other tasks to participate in. The expected secrecy rate for Alice and Bob is given as follows:

$$\text{CS} = \sum_{i=1}^{n} \left(\ln \left(1 + \frac{h_i P_i}{\sigma^2} \right) - q_i \ln \left(1 + \frac{h_{Ei} P_i}{\sigma_E^2} \right) \right), \tag{1}$$

where $\boldsymbol{P} = (P_1, \ldots, P_n)$ with P_i is the power transmitted by Alice through channel i, $\sum_{i=1}^{n} P_i = \overline{P}$, and $\overline{P}$ is the total power budget to transmit, σ^2 and σ_E^2 are background noise of the main and eavesdropper's channels. We assume that $h_i/\sigma^2 \geq h_{Ei}/\sigma_E^2$ for any i. This relation ensures that we are dealing with the case where Alice-Bob's channel is more capable than the eavesdropping channel. Let Π be the set of all feasible strategies for Alice. Alice can control the power allocation $\boldsymbol{P}$ of the signal among the channels to transmit, i.e. a strategy for Alice is a power vector $\boldsymbol{P}$.

The (secrecy) payoff to Alice is given by the (expected) secrecy capacity:

$$v_A(\boldsymbol{P}) = \sum_{i=1}^{n} \left(\ln \left(1 + h_i P_i \right) - q_i \ln \left(1 + h_{Ei} P_i \right) \right), \tag{2}$$

where for simplification, we use the following notation: $h_i := h_i/\sigma^2$ and $h_{Ei} := h_{Ei}/\sigma_E^2$ for any i.

We shall say that the communication operates in the *secrecy mode* if Alice wants to maximize this payoff, i.e., to solve the following optimization problem:

$$\boldsymbol{P}_A = \arg\max_{\boldsymbol{P} \in \Pi} v_A(\boldsymbol{P}),$$

This is a generalization of classical secrecy communication problem (see, for example, [15,21]) for the case where there is no complete information whether the channels are eavesdropped or not. Solution to this problem is given in the following theorem.

Theorem 1. *The optimal strategy $\boldsymbol{P}_A$ in secrecy mode is given as follows*

$$P_{Ai} = P_{Ai}(\omega) := \frac{1}{2h_i h_{Ei}} \left[\frac{(1-q_i)h_i h_{Ei}}{\omega} - h_i - h_{Ei} \right.$$

$$\left. + \sqrt{(h_i - h_{Ei})^2 + \frac{2(1+q_i)(h_i - h_{Ei})h_i h_{Ei}}{\omega} + \left(\frac{(1-q_i)h_i h_{Ei}}{\omega} \right)^2} \right]_+ \tag{3}$$

for $i \in \{1, \ldots, n\}$ where $\omega = \omega_A$ is given as the unique root in $(0, \delta_{\max})$ with $\delta_{\max} = \max_i \{h_i - h_{Ei}\}$ of the water-filling equation:

$$F(\omega_A) := \sum_{i=1}^{n} P_{Ai}(\omega_A) = \overline{P}. \tag{4}$$

By (3), $F(\omega)$ is continuous for $\omega > 0$, decreasing in $(0, \delta_{\max})$ such that $F(\delta_{\max}) = 0$ and $F(\omega)$ tends to infinity while ω tends to zero. Thus, the root of the Eq. (4) can be found by the bisection method.

3   Security Mode

The secrecy capacity given by (1) reflects the rate at which communication received by Bob from Alice can be properly coded (i.e., privacy amplification) to be confidential when eavesdropped by Eve. The metric,

$$L(\boldsymbol{P}) := \sum_{i=1}^{n} q_i \ln\left(1 + h_{Ei} P_i\right)$$

reflects the probable amount of leakage that privacy amplification would have to cope with. This probable leakage, though, can also be viewed in other ways, such as providing evidence that there is a communication session between Alice and Bob (i.e. the security notion of detectability), or the amount to which the communication rate was hindered because of the adversary eavesdropping. Consequently, this metric captures a security risk associated with the communication between Alice and Bob. Thus, maintaining secure communication, Alice faces a *dual* objective problem:

- *The first objective (secrecy)* is to maximize the rate at which information confidentially can be shared between Alice and Bob;
- *The second objective (security)* is to minimize probable information leakage during communication between Alice and Bob.

In this paper, in order to be able to cast our analysis in the framework of concave programming, we consider the expected eavesdropped SNR (Signal-to-Noise Ratio) by Eve as the security objective, i.e.,

$$v_E(\boldsymbol{P}) := \text{SNR}_E(\boldsymbol{P}) = \sum_{i=1}^{n} q_i h_{Ei} P_i.$$

As examples of scenarios where SNR has been considered as an objective, see [1] for selfish and cooperative transmission protocols and [30] for anti-jamming strategies.

Note that, although Eve is a passive agent, this v_E can be considered either as a payoff to Eve or as a (security) cost to Alice. When Alice aims to minimize her security cost, we shall call such a mode of communication as being in a *security mode*. Then, in this case, the security cost v_E is multiplied by -1, i.e., $v_{AE}(\boldsymbol{P}) = -v_E(\boldsymbol{P})$, and can be considered as a security objective that Alice wants to maximize in security mode. The strategy $\boldsymbol{P}_{AE}$, maximizing security objective to Alice, focuses all the power budget on channels I_E with the minimal expected eavesdropping channel gains i.e.,

$$P_{AEi} \begin{cases} \geq 0, & i \in I_E, \\ = 0, & i \notin I_E, \end{cases}$$

where $I_E = \{i : q_i h_{Ei} = \min_j q_j h_{Ej}\}$.

This dual objective problem puts Alice on the horns of a dilemma.

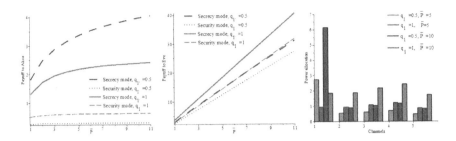

Fig. 2. Payoff to Alice (left), and payoff to Eve (middle) as functions on total power budget $\overline{P}$, and the optimal strategy $\boldsymbol{P}_A$ in secrecy mode for $\overline{P} = 5, 10$.

- *On the one hand*, an increase in the secrecy rate received by Bob (i.e., in communication secrecy) might also yield a gain in what was eavesdropped by Eve, and, thus, in a reduction of security.
- *On the other hand*, a decrease in what was eavesdropped by Eve (and, thus, an improvement in communication security) could also yield a decrease in the throughput received by Bob (i.e., in a decrease in the secrecy rate).

This dilemma is illustrated by Fig. 2 by depicting the payoffs to Eve and Alice as functions of the transmission power $\overline{P}$ with $n = 5$, $h = (9, 6.5, 5, 5.5, 6)$, $h_E = (5, 4, 3, 2.9, 4)$ and $q = (x, 1, 1, 1, 1)$ with $x = 0.5, 1$. An increase in the power budget leads to an increase in payoffs to Alice and Eve. An increase in the probability that a channel might be eavesdropped yields a decrease in Alice's payoff and in an increase in Eve's payoff. For $q_1 = 0.5$, channel 1 with $q_1 h_{E1} = 2.5$ has the minimal expected eavesdropping gain, and so in security mode Alice uses only this channel for communication. While, for $q_1 = 1$, channel 4 with $q_4 h_{E4} = 2.9$ has the minimal expected eavesdropping gain, and Alice switches to use it in security mode. In secrecy mode, Alice uses all of the channels focusing the essential part of her effort on channel 1 for $q_1 = 0.5$.

4   Secrecy Communication with Fixed Security Level

To solve the dilemma between secrecy and security, we introduce the notion of priority for fulfilment of the objectives. We assume that Alice wants to communicate secretly with Bob in such a way that Eve cannot gain more than Alice allows her to. We shall denote by ϵ_E the threshold value for the Eve's payoff. Thus, Alice wants to solve the following optimization problem:

$$\max v_A(\boldsymbol{P}),$$
$$\text{subject to } v_E(\boldsymbol{P}) \leq \epsilon_E, \boldsymbol{P} \in \Pi. \tag{5}$$

Since, the condition $v_E(\boldsymbol{P}) \leq \epsilon_E$ is equivalent to $v_{AE}(\boldsymbol{P}) \geq -\epsilon_E$, this $-\epsilon_E$ can be considered as a *security level* for the Alice-Bob communication. Smaller ϵ_E means a higher level of security and so smaller leakage. While larger ϵ_E means a smaller level of security, and, so, higher information leakage.

If Alice wants to communicate with Bob according to (5), we shall call such a communication as operating in the *secrecy mode with a fixed security level*. Note that, (5) can be considered as a generalization of the classical secrecy problem. Of course, for such a problem the set of feasible strategies Π_E has to be non-empty, where

$$\Pi_E := \left\{ \boldsymbol{P} \in \Pi : \sum_{i=1}^{n} q_i h_{Ei} P_i \le \epsilon_E \right\}.$$

Since v_E is linear in $\boldsymbol{P}$, the following upper and lower bounds for v_E holds

$$E_{\min} \le v_E(\boldsymbol{P}) \le E_{\max},$$

where $E_{\min} = \overline{P} \min_i q_i h_{Ei}$ and $E_{\max} = \overline{P} \max_i q_i h_{Ei}$.

Thus, the set of feasible strategies Π_E is not empty if and only if

$$E_{\min} \le \epsilon_E. \tag{6}$$

Theorem 2. *Let Alice communicate in the secrecy mode with a fixed security level. Further, let the set of feasible strategies be non empty, i.e., (6) holds. Then, there is a unique optimal strategy $\boldsymbol{P}_{\epsilon_E}$ for Alice.*

(a) Let

$$\epsilon_{E\,\max} := \sum_{i=1}^{n} q_i h_{Ei} P_{Ai} \le \epsilon_E \tag{7}$$

with $\boldsymbol{P}_A$ given by Theorem 1. Then $\boldsymbol{P}_{\epsilon_E} = \boldsymbol{P}_A$.
(b) Let

$$E_{\min} < \epsilon_E < \epsilon_{E\,\max}. \tag{8}$$

Then

$$P_{\epsilon_E i} = P_{\epsilon_E i}(\omega, \nu) := \frac{1}{2 h_i h_{Ei}} \left[\frac{(1 - q_i) h_i h_{Ei}}{\omega + \nu h_{Ei}} - h_i - h_{Ei} \right.$$

$$\left. + \sqrt{(h_i - h_{Ei})^2 + \frac{2(1 + q_i)(h_i - h_{Ei}) h_i h_{Ei}}{\omega + \nu h_{Ei}} + \left(\frac{(1 - q_i) h_i h_{Ei}}{\omega + \nu h_{Ei}} \right)^2} \right]_{+}$$

$$\tag{9}$$

for $i \in \{1, \ldots, n\}$, where ω and ν is the unique solution to the following two equations:

$$F(\omega, \nu) := \sum_{i=1}^{n} P_{\epsilon_E i}(\omega, \nu) = \overline{P} \tag{10}$$

and

$$F_E(\omega, \nu) := \sum_{i=1}^{n} q_i P_{\epsilon_E i}(\omega, \nu) = \epsilon_E. \tag{11}$$

(c) Let

$$\epsilon_E = E_{\min}. \tag{12}$$

Then

$$\boldsymbol{P}_{\epsilon_E i} = \boldsymbol{P}_{E \min i}(\omega) := \begin{cases} P_{Ai}(\omega), & i \in I_E, \\ 0, & i \notin I_E, \end{cases}$$

i.e., $\boldsymbol{P}_{\epsilon_E}$ is zero outside of I_E, and it is a restriction of $\boldsymbol{P}_A$ in I_E with $\omega = \omega_{\min}$ is the unique root of the equation

$$\sum_{i \in I_E} P_{Ai}(\omega_{\min}) = \overline{P}.$$

Note that, the condition (10) reflects the fact that ω and ν have to be such that $\boldsymbol{P}_{\epsilon_E}(\omega, \nu)$, given by (9), is a strategy involving allocating the total power $\overline{P}$ among all the channels. (11) reflects that ω and ν have to be such that $\boldsymbol{P}_{\epsilon_E}(\omega, \nu)$ is a boundary point for the set of feasible strategies Π_E.

Condition (7) means that, if the optimal solution $\boldsymbol{P}_A$ for the secrecy mode is also feasible for the secrecy mode with fixed security level, then this solution is also optimal for that mode.

5   Iterative Algorithm for Finding the Optimal Strategy

To design an iterative algorithm that finds the optimal ω and ν given by Theorem 2 (which return the optimal strategy $\boldsymbol{P}_{\epsilon_E}$) we first describe some properties of the optimal solution.

(i) By Theorem 1, $F(\omega_A, 0) = \overline{P}$.

(ii) Due to $F(0, \nu)$ being decreasing from infinity (when ν is near 0) to zero for $\nu = \Delta_{\max} := ((h_i - h_{Ei})/h_{Ei})$, there exists the unique ν_A in $(0, \Delta_{\max})$ such that $F(0, \nu_A) = \overline{P}$.

(iii) Since $F(\omega, \nu)$ is decreasing in ω and ν, by (i) and (ii), for any fixed $\nu \in [0, \nu_A]$ there is a unique $\Omega(\nu)$ such that $F(\Omega(\nu), \nu) = \overline{P}$. Also, $\Omega(\nu)$ is continuous and decreasing from ω_A for $\nu = 0$ to zero for $\nu = \nu_A$.

(iv) Due to $F_E(\omega, 0)$ is decreasing from infinity (when ω is near 0) to zero for $\omega = \delta_{\max}$, there is the unique $\omega_E \in (0, \delta_{\max})$ such that $F_E(\omega_E, 0) = \epsilon_E$.

(v) Due to $F_E(0, \nu)$ is decreasing from infinity (when ν is near 0) to zero for $\nu = \Delta_{\max}$, there is the unique $\nu_E \in (0, \Delta_{\max})$ such that $F_E(0, \nu_E) = \epsilon_E$.

(vi) Since $F_E(\omega, \nu)$ is decreasing in ω and ν, by (iv) and (v), for any $\omega \in [0, \omega_E]$ there is a function $N(\omega)$ such that $F_E(\omega, N(\omega)) = \epsilon_E$. Also, $N(\omega)$ is continuous and decreasing from ν_E for $\omega = 0$ to zero for $\omega = \omega_E$.

By (i)–(iii), we have that the condition (7) is equivalent to $F_E(\omega_A, 0) \geq \epsilon_E$. Similarly, by (iv)–(vi), due to F_E is decreasing in ω, we have that the condition (7) is equivalent to $\omega_A \leq \overline{\omega}_*$, and the condition (8) is equivalent to $\omega_A > \omega_E$. Thus, the optimal strategy $\boldsymbol{P}_{\epsilon_E}$ also can be described as follows:

$$\boldsymbol{P}_{\epsilon_E} = \begin{cases} \boldsymbol{P}(\omega_A, 0), & \omega_A \leq \omega_E, \\ \boldsymbol{P}_{\epsilon_E}(\omega, \nu) \text{ with } \omega = \Omega(\nu), \nu = N(\omega), & \omega_A > \omega_E. \end{cases}$$

By Theorem 2, the intersection of the curves $\Omega(\nu)$ and $N(\omega)$ is uniquely defined, and the optimal ν is given as a fixed point of $\nu = N(\Omega(\nu))$. Then, since the functions $\Omega(\nu)$ and $N(\omega)$ are monotonic decreasing, the iterative algorithm

$$\nu_{i+1} = N(\Omega(\nu_i)) \text{ for } i = 0, 1, \ldots$$

converges to the fixed point. This approach to find the optimal strategy is described in detail in Algorithm 1.

Algorithm 1. Finding the optimal strategy $\boldsymbol{P}_{\epsilon_E}$.

Solve $F(\omega_A, 0) = \overline{P}$ for $\omega_A \in (0, \delta_{\max}]$ by bisection method
Solve $F(0, \nu_A) = \overline{P}$ for $\nu_A \in (0, \Delta_{\max})$ by bisection method
Solve $F_E(\omega_E, 0) = \epsilon_E \ \omega_E \in (0, \delta_{\max}]$ by bisection method
Solve $F_E(0, \nu_E) = \epsilon_E$ for $\nu_E \in (0, \Delta_{\max})$ by bisection method
if $\omega_A \leq \omega_E$ **then**
   return strategy $\boldsymbol{P}_{\epsilon_E}(\omega_A, 0)$
else
   let $i = 0$
   let $\nu_0 = \nu_E$
   Find $\Omega(\nu_0)$ in $[0, \omega_A]$ by bisection method such that $F(\Omega(\nu_0), \nu_0) = \overline{P}$
   let $\omega_0 = \Omega(\nu_0)$
   repeat
     $i = i + 1$
     Find $N(\omega_i)$ in $[0, \nu_A]$ by bisection method such that $F_E(\omega_i, N(\omega_i)) = \epsilon_E$
     let $\nu_{i+1} = N(\omega_i)$
     Find $\Omega(\nu_i)$ in $[0, \omega_*]$ by bisection method such that $F(\Omega(\nu_i), \nu_i) = \overline{P}$
     let $\omega_{i+1} = \Omega(\nu_i)$
   until $|\nu_{i+1} - \nu_i| > \theta$ (θ is tolerance of the algorithm)
   return strategy $\boldsymbol{P}_{\epsilon_E}(\omega_{i+1}, \nu_i)$
end if

Figure 3(a) illustrates the convergence of the iterative algorithm. Figures 3(b) and (c) illustrate that an increase in the threshold value ϵ_E yields an increase in

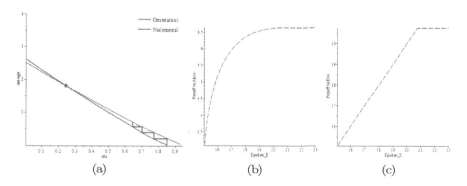

     (a)   (b)   (c)

Fig. 3. (a) Convergence of the iterative algorithm, (b) payoff to Alice and (c) payoff to Eve as functions of ϵ_E.

the payoff to Alice as well as in the payoff to Eve. Thus, *a problem arises:* what bargaining ϵ_E has to be to fulfil two objectives: to maintain secrecy communication (i.e., to look for an increase in Alice's payoff) and communication security (i.e., to look for a decrease in Eve's payoff).

6   Bargaining over Security Level

The bargaining solution is based on the concept of negotiation between *agents* to get a solution that is appropriate for both of them, and it is obtained on the basis of an axiomatic approach [26]:

First, to formulate the bargaining problem, we have to define *agents* who are bargaining. We consider two objectives for Alice's communication (secrecy and security) as such agents. Thus, we deal with a two agent bargaining problem.

Second, we have to define a pair $(G, \boldsymbol{g}_0)$. Here G is so called *feasibility* set, which is a subset of $\mathbb{R}^2$. Its elements reflect all possible payoffs that these two agents could get if they achieve an agreement. $\boldsymbol{g}_0 = (\boldsymbol{g}_{01}, \boldsymbol{g}_{02})$ is a *disagreement* point, where g_{01} and g_{02} are payoffs to the agents if they do not achieve an agreement.

Due to payoff for the secrecy objective is v_A, and payoff for the security objective is v_{AE}, we can define in plane (v_{AE}, v_A) all the payoffs of these objectives as a parameterized curve on ϵ_E as follows:

$$G = \{(v_{AE}(\boldsymbol{P}_{\epsilon_E}), v_A(\boldsymbol{P}_{\epsilon_E})) : \epsilon_E \in [E_{\min}, \epsilon_{E\,\max}]\}.$$

Since $v_A(\boldsymbol{P}_{\epsilon_E})$ is the unique solution of the concave optimization problem (5) and $v_{AE}(\boldsymbol{P}_{\epsilon_E}) = -\epsilon_E$ for $\epsilon_E \in [E_{\min}, \epsilon_{E\,\max}]$, G is a concave curve in plane (v_{AE}, v_A). Thus, each point of this G is Pareto optimal. Recall that point $(g_1, g_2) \in G$ is said to be Pareto optimal, if and only if there is no $(g_1', g_2') \in G$ such that $g_i' \geq g_i$ for $i = 1, 2$ and $g_i' > g_i$ for at least one i. Thus, G can be considered as a feasibility set for the bargaining problem. As the disagreement point we can consider:

$$(v_{AEd}, v_{Ad}) = (-\epsilon_{E\,\max}, v_A(\boldsymbol{P}_{\epsilon_{E\,\max}})) = (-\epsilon_{E\,\max}, v_A(\boldsymbol{P}_A)).$$

Third, to design the bargaining solution, we have to make a decision on the axiomatic approach we are going to employ, and which define the properties of the bargaining solution. In this paper, as the bargaining solution, we consider the Kalai-Smorodinsky bargaining solution (KSBS), which is aimed at preserving the ratio of maximum gains for the agents. In [17], it was shown that the KSBS is the Pareto-optimal point on the line joining the disagreement point $\boldsymbol{d}_0$ with the so-called *utopia point* $\boldsymbol{g}_u = (g_{u1}, g_{u2})$, elements of which g_{u1} and g_{u2} are the maximal possible payoff to agent 1 and agent 2. The utopia point for the considered model is

$$(v_{AEu}, v_{Au}) = (-E_{\min}, v_A(\boldsymbol{P}_{E_{\min}})).$$

Forth, we design the KSBS as the intersection of G and the line L where

$$v_A = L(v_{AE}) := \frac{v_{AE} + \epsilon_{E\,\max}}{\epsilon_{E\,\max} - E_{\min}}(v_A(\boldsymbol{P}_{E_{\min}}) - v_A(\boldsymbol{P}_A)) + v_A(\boldsymbol{P}_{E_{\min}}).$$

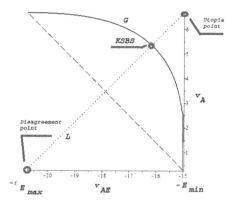

Fig. 4. The payoff/cost curve with the bargaining point for $q_1 = 1$, $\overline{P} = 10$.

This intersection can be found by Algorithm 2 using the bisection method (Fig. 4).

In the considered example, the utopia point is $(-15.0, 0.295)$ and the disagreement point is $(-20.748, 6.634)$. Thus, the Kalai-Smorodinsky bargaining solution is $(-16.20, 5.37)$ with $\epsilon_E = 16.20$. Figure 5(a) illustrates that an increase in the power budget as well as in the probability to be eavesdropped yield an increase in the bargaining value of ϵ_E. Figures 5(b) and (c) illustrate that the bargaining values for payoffs to Alice and to Eve are trade-off values for the ones returned by optimal strategies in security mode and secrecy mode.

Algorithm 2. Bargaining value for security level.

let $\epsilon_a = \epsilon_{E\,min}$, $\epsilon_b = \epsilon_{E\,max}$
repeat
    let $\epsilon = (\epsilon_a + \epsilon_b)/2$
    if $v_A(P_\epsilon) > L(\epsilon)$ **then** $\epsilon_b = \epsilon$ **else** $\epsilon_a = \epsilon$
until $\epsilon_b - \epsilon_a > \theta$ (θ is tolerance of the algorithm)
return ϵ

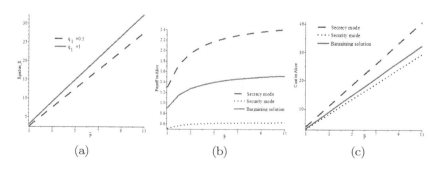

Fig. 5. (a) The bargaining value of ϵ_E, (b) the bargaining value of payoff to Alice and (c) cost to Alice as functions of $\overline{P}$ for $q_1 = 1$.

7   Conclusions

In this paper, a new form of secure communication problem was formulated involving two objectives (namely, secrecy and security). As the objective for secrecy communication we used the secrecy rate, while the probable leakage in the communication was used as the objective for communication security. We first showed that the solution of such a dual objective problem might present a dilemma since the optimal solution for one of them might not be optimal for the other. To solve this dilemma, we generalized a classical problem of secret communication by introducing a notion of security level for communication, and solved it in closed form for a fixed security level. *Then*, we found the optimal security level as a Kalai-Smorodinsky bargaining solution, which maintains the ratios of maximal gains for cooperative fulfilment of the both objectives.

8   Appendix

8.1   Proof of Theorem 1

Since $v_A(\boldsymbol{P})$ is strictly concave, there is the unique optimal $\boldsymbol{P}$. To find it we define the Lagrangian as follows:

$$L_\omega(\boldsymbol{P}) = \sum_{i=1}^{n} (\ln(1 + h_i P_i) - q_i \ln(1 + h_{Ei} P_i)) + \omega(\overline{P} - \sum_{i=1}^{n} P_i),$$

where ω is Lagrange multiplier. Thus, the optimal $\boldsymbol{P}$ is given by the Karush-Kuhn-Tucker (KKT) conditions:

$$\frac{h_i}{1 + h_i P_i} - \frac{q_i h_{Ei}}{1 + h_{Ei} P_i} - \omega \begin{cases} = 0, & P_i > 0, \\ \leq 0, & P_i = 0. \end{cases} \tag{13}$$

Then, by (13), $\omega > 0$. Solving (13) yields that $\boldsymbol{P}$ has to be given by (3). Taking into account that $\boldsymbol{P} \in \Pi$ implies (4), and the result follows.   ∎

8.2   Proof of Theorem 2

Since $v_A(\boldsymbol{P})$ is strictly concave, there is a unique optimal $\boldsymbol{P}$. To find it, we define the Lagrangian as follows:

$$L_{\omega,\nu}(\boldsymbol{P}) = \sum_{i=1}^{n} (\ln(1 + h_i P_i) - q_i \ln(1 + h_{Ei} P_i))$$

$$+ \nu(\epsilon_E - \sum_{i=1}^{n} q_i h_{Ei} P_i) + \omega(\overline{P} - \sum_{i=1}^{n} P_i),$$

where $\nu \geq 0$ and ω are Lagrange multipliers. Thus, the optimal $\boldsymbol{P}$ is given by the Karush-Kuhn-Tucker (KKT) conditions:

$$\frac{h_i}{1 + h_i P_i} - \frac{q_i h_{Ei}}{1 + h_{Ei} P_i} - q_i h_{Ei} \nu - \omega \begin{cases} = 0, & P_i > 0, \\ \leq 0, & P_i = 0, \end{cases} \tag{14}$$

$$0 = \nu \left(\epsilon_E - \sum_{i=1}^{n} q_i h_{Ei} P_i \right) \text{ and } 0 = \omega \left(\overline{P} - \sum_{i=1}^{n} P_i \right). \tag{15}$$

Since this is a differentiable, convex optimization problem with linear constraints, not only are the KKT conditions mentioned above sufficient, they are also necessary for optimality. Two cases might arise for the optimal $\boldsymbol{P}$:

(i) $\sum_{i=1}^{n} q_i h_{Ei} P_i < \epsilon_E$,

(ii) $\sum_{i=1}^{n} q_i h_{Ei} P_i = \epsilon_E$.

Let (i) hold. Then, by (15), $\nu = 0$. Thus, by (14), $\boldsymbol{P}$ has to be given by Theorem 1. Then, by (i), (7) also has to hold, and (a) follows.

Let (ii) hold. Then, by (15), $\nu \geq 0$. Thus, by (14), $\boldsymbol{P}$ has to be given by (9). Substituting this $\boldsymbol{P}$ into (ii) and taking into account that $\boldsymbol{P} \in \Pi$ imply (10) and (11), and the result follows. ∎

References

1. Altman, E., Avrachenkov, K., Garnaev, A.: Transmission power control game with SINR as objective function. In: Altman, E., Chaintreau, A. (eds.) NET-COOP 2008. LNCS, vol. 5425, pp. 112–120. Springer, Heidelberg (2009). doi:10.1007/978-3-642-00393-6_14

2. Alvim, M.S., Chatzikokolakis, K., Palamidessi, C., Smith, G.: Measuring information leakage using generalized gain functions. In: 25th IEEE Computer Security Foundations Symposium, pp. 265–279 (2012)

3. Balmahoon, R., Cheng, L.: Information leakage of heterogeneous encoded correlated sequences over an eavesdropped channel. In: IEEE International Symposium on Information Theory (ISIT), pp. 2949–2953 (2015)

4. Bashar, S., Ding, Z.: Optimum power allocation against information leakage in wireless network. In: IEEE Global Telecommunications Conference (GLOBECOM), pp. 1–6 (2009)

5. Csiszár, I., Körner, J.: Broadcast channels with confidential messages. IEEE Trans. Inf. Theory **24**, 339–348 (1978)

6. Garnaev, A., Baykal-Gursoy, M., Poor, H.V.: Incorporating attack-type uncertainty into network protection. IEEE Trans. Inf. Forensics Secur. **9**, 1278–1287 (2014)

7. Garnaev, A., Baykal-Gursoy, M., Poor, H.V.: A game theoretic analysis of secret and reliable communication with active and passive adversarial modes. IEEE Trans. Wirel. Commun. **15**, 2155–2163 (2016)

8. Garnaev, A., Trappe, W.: The eavesdropping and jamming dilemma in multi-channel communications. In: IEEE International Conference on Communications (ICC), pp. 2160–2164 (2013)

9. Garnaev, A., Trappe, W.: Secret communication when the eavesdropper might be an active adversary. In: Jonsson, M., Vinel, A., Bellalta, B., Belyaev, E. (eds.) MACOM 2014. LNCS, vol. 8715, pp. 121–136. Springer, Heidelberg (2014). doi:10.1007/978-3-319-10262-7_12

10. Garnaev, A., Trappe, W.: To eavesdrop or jam, that is the question. In: Mellouk, A., Sherif, M.H., Li, J., Bellavista, P. (eds.) ADHOCNETS 2013. LNICSSITE, vol. 129, pp. 146–161. Springer, Heidelberg (2014). doi:10.1007/978-3-319-04105-6_10

11. Garnaev, A., Trappe, W.: Anti-jamming strategies: a stochastic game approach. In: Agüero, R., Zinner, T., Goleva, R., Timm-Giel, A., Tran-Gia, P. (eds.) MONAMI 2014. LNICSSITE, vol. 141, pp. 230–243. Springer, Heidelberg (2015). doi:10.1007/978-3-319-16292-8_17

12. Garnaev, A., Trappe, W.: One-time spectrum coexistence in dynamic spectrum access when the secondary user may be malicious. IEEE Trans. Inf. Forensics Secur. **10**, 1064–1075 (2015)

13. Garnaev, A., Trappe, W.: A bandwidth monitoring strategy under uncertainty of the adversary's activity. IEEE Trans. Inf. Forensics Secur. **11**, 837–849 (2016)

14. Garnaev, A., Trappe, W., Kung, C.-T.: Optimizing scanning strategies: selecting scanning bandwidth in adversarial RF environments. In: 8th International Conference on Cognitive Radio Oriented Wireless Networks (CROWNCOM), pp. 148–153 (2013)

15. Gopala, P.K., Lai, L., El Gamal, H.: On the secrecy capacity of fading channels. IEEE Trans. Inf. Theory **54**, 4687–4698 (2008)

16. Ho, Z., Jorswieck, E., Engelmann, S.: Information leakage neutralization for the multi-antenna non-regenerative relay-assisted multi-carrier interference channel. IEEE J. Sel. Areas Commun. **31**, 1672–1686 (2013)

17. Kalai, E., Smorodinsky, M.: Other solutions to Nash's bargaining problem. Econometrica **43**, 513–518 (1975)

18. Kim, M., Hwang, E., Kim, J.: Analysis of eavesdropping attack in mmWave-based WPANs with directional antennas. J. Wirel. Netw. (2015). doi:10.1007/s11276-015-1160-4

19. Koyluoglu, O.O., El Gamal, H., Lai, L., Poor, H.V.: Interference alignment for secrecy. IEEE Trans. Inf. Theory **57**, 3323–3332 (2011)

20. Li, Z., Yates, R., Trappe, W.: Secure communication with a fading eavesdropper channel. In: IEEE International Symposium on Information Theory (ISIT), pp. 1296–1300 (2007)

21. Liang, Y., Poor, H.V., Shamai, S.: Information theoretic security. Found. Trends Commun. Inf. Theory **5**(4–5), 355–580 (2009)

22. Liang, Y., Poor, H.V., Shamai, S.: Secure communications over fading channels. IEEE Trans. Inf. Theory **54**, 2470–2492 (2008)

23. Lin, M., Ge, J., Yang, Y.: An effective secure transmission scheme for AF relay networks with two-hop information leakage. IEEE Commun. Lett. **17**, 1676–1679 (2013)

24. Mukherjee, A., Fakoorian, S.A.A., Huang, J., Swindlehurst, A.L.: Principles of physical layer security in multiuser wireless networks: a survey. IEEE Commun. Surv. Tutor. **16**, 1550–1573 (2014)

25. Negi, R., Goel, S.: Secret communication using artificial noise. In: 62nd IEEE Vehicular Technology Conference (VTC), pp. 1906–1910 (2005)

26. Owen, G.: Game Theory. Academic Press, New York (1982)

27. Wu, Y., Liu, K.J.R.: An information secrecy game in cognitive radio networks. IEEE Trans. Inf. Forensics Secur. **6**, 831–842 (2011)

28. Xiong, Q., Liang, Y.-C., Li, K.H., Gong, Y.: An energy-ratio-based approach for detecting pilot spoofing attack in multiple-antenna systems. IEEE Trans. Inf. Forensics Secur. **10**, 932–940 (2015)

29. Yang, C., Li, J., Anpalagan, A.: Strategic bargaining in wireless networks: basics, opportunities and challenges. IET Commun. **8**, 3435–3450 (2014)

30. Yang, D., Zhang, J., Fang, X., Richa, A., Xue, G.: Optimal transmission power control in the presence of a smart jammer. In: IEEE Global Communications Conference (GLOBECOM), pp. 5506–5511 (2012)
31. Zhu, Q., Saad, W., Han, Z., Poor, H.V., Basar, T.: Eavesdropping and jamming in next-generation wireless networks: a game-theoretic approach. In: IEEE Military Communications Conference (MILCOM), pp. 119–124 (2011)

An Analytical Model for Perpetual Network Codes in Packet Erasure Channels

Peyman Pahlevani[1]([⊠]), Sérgio Crisóstomo[2], and Daniel E. Lucani[3]

[1] Department of Computer Science and Information Technology,
Institute for Advanced Studies in Basic Sciences (IASBS), Zanjan, Iran
`pahlevani@iasbs.ac.ir`
[2] Instituto de Telecomunicações and Faculdade de Ciências,
Universidade do Porto, Porto, Portugal
`slc@dcc.fc.up.pt`
[3] Department of Electronic Systems, Aalborg University, Aalborg, Denmark
`del@es.aau.dk`

Abstract. Perpetual codes provide a sparse, but structured coding for fast encoding and decoding. In this work, we illustrate that perpetual codes introduce linear dependent packet transmissions in the presence of an erasure channel. We demonstrate that the number of linear dependent packet transmissions is highly dependent on a parameter called the *width* (ω), which represents the number of consecutive non-zero coding coefficient present in each coded packet after a pivot element. We provide a mathematical analysis based on the width of the coding vector for the number of transmitted packets and validate it with simulation results. The simulations show that for $\omega = 5$, generation size $g = 256$, and low erasure probability on the link, a destination can receive up to 70% overhead in average. Moreover, increasing the *width*, the overhead contracts, and for $\omega \geq 60$ it becomes negligible.

1 Introduction

Network Coding (NC) is a promising paradigm that has been shown to provide benefits in many different applications from storage to computer communication. NC allows each node in a communication network to code packets together, as opposed to other rateless codes (e.g., LT codes [6], Raptor codes [10]), where only the source and destination nodes can perform coding/decoding operations. In the early work by Alhswede et al., it was shown that, in a multicast network, the maximum capacity can be achieved using network coding [1]. Li et al. [5] have shown that linear network codes are sufficient to achieve the multicast capacity when transmitting from a single source to multiple receivers. Random Linear Network Coding (RLNC) [4] provides a practical and decentralized approach to exploit network coding in highly dynamic and decentralized environments, ranging from core Internet networks to wireless mesh networks. For changing channel conditions, RLNC reduces the signaling overhead and increases robustness, when compared to standard routing protocols. However, the complexity

© Springer International Publishing AG 2016
T.K. Madsen et al. (Eds.): MACOM 2016, LNCS 10121, pp. 126–135, 2016.
DOI: 10.1007/978-3-319-51376-8_10

of encoding and decoding in RLNC is a barrier to exploit on computationally restricted platforms (e.g., sensors [8], mobile phones [3]), which are set to have an ever growing share of Internet connected devices due to the new trend of the Internet of Things and the ubiquity of mobile phones.

Recently, Heide et al. [2] exploited an unpublished paper entitled "Perpetual Codes: Cache-friendly decoding" [7], with the aim of reducing RLNC decoding complexity. Perpetual codes advocate the use of a sparse and structured coding of the packets, which allows both encoders and decoders to process data faster in real processors. Perpetual codes also allow for an efficient representation of the coding coefficients, which signals which packets were combined to form a coded packet. The structure and density of the coded packets can be retained if recoding is performed judiciously.

In this paper we provide a model to quantify the performance of perpetual codes over packet erasure channels. Our model shows that the performance is tightly dependent on the *width* of the coding vectors. Moreover, our analysis illustrate that, when transmitting over an erasure channel with erasure probability $\epsilon = 0.2$, using generation size $g = 256$, and $width = 5$, the linearly dependent packets reception represents 70% of the total received packets. The reception of linearly dependent packets is strongly attenuated by increasing the *width* parameter, becoming negligible for $width \geq 60$.

The paper is organized as follows. In Sect. 2 we introduce the perpetual codes and the motivation for this work. In Sect. 3 we provide a model for the analysis of perpetual codes. Numerical results are presented in Sect. 4. Finally, a conclusion follows in Sect. 5.

2   Perpetual Codes

For this study, we assume that data is transmitted from a source node to a destination, divided into generations of size g. Each generation is composed by coded packets that are a linear combination of the source packets. The source creates the coded packets and transmits the coded packet until the receiver decodes the generation. Whenever the receiver gets enough Degree of Freedom (DOF) to decode a generation, it transmits an acknowledgment to the sender to make it stop from creating new coded packets.

The transmitted coded packets can be lost due errors on the channel. The packet erasure probability between the source and the destination is ϵ. However, we assume that the channel from the destination to the source is perfect. Thus, the acknowledgment packets are received immediately after they are created.

The coding coefficients of a coded packet are chosen over a finite field F_q of size q. The group of coding coefficients in a coded packet is called the coding vector. A coding vector in perpetual codes is composed by a pivot element that is a coefficient with value 1, followed by ω coefficients drawn at random from the field F_q. The remaining coefficients are set to zero. We call ω the *width* of perpetual codes. This is a parameter set at the source and maintained constant during the entire transmission process.

There can be two different strategies to transmit the coded packets in perpetual codes.

Random Strategy: The pivot elements are chosen randomly. The corresponding coding vectors are used to generate coded packets to be transmitted. The sender stops the transmission process when it receives an acknowledgment from the destination.

Systematic Strategy: For a generation of size g, a matrix with g coding vectors is created at the sender, as shown in Fig. 1. The elements of this matrix are chosen sequentially, from the first to the last row, in order to generate the coded packets to be transmitted. Figure 1 shows the coding matrix for a generation with size eight, w equal to three, and $\alpha_{i,j} \in F_q$. In the presence of packet losses (packet erasure channel), the sender may have to perform multiple sequential transmission rounds in a round robin fashion, until the destination is able to decode the full generation. In each round the coefficients from the coding matrix are replaced with new values. The sender stops the transmission process when it receives an acknowledgment from the destination. In this paper we only consider the systematic strategy in the choice of pivot elements.

To understand the effect of linearly dependent packets, we define a parameter called *overhead* as:

$$O = \frac{R_d}{R_i}, \tag{1}$$

where, R_d corresponds to the number of linear dependent packets and R_i is the number linear independent packets at the destination.

Let us consider the example of Fig. 2 where the sender must transmit a generation built from seven source packets $\{P_1, P_2, \cdots, P_7\}$ using $w = 1$. As shown in this figure, the first and the second coded packets are $P_1 + \alpha_{1,2} \cdot P_2$ and $P_2 + \alpha_{2,3} \cdot P_3$. The sender keeps creating and transmitting coded packets until

$$\omega$$

1	$\alpha_{1,2}$	$\alpha_{1,3}$	$\alpha_{1,4}$	0	0	0	0
0	1	$\alpha_{2,3}$	$\alpha_{2,4}$	$\alpha_{2,5}$	0	0	0
0	0	1	$\alpha_{3,4}$	$\alpha_{3,5}$	$\alpha_{3,6}$	0	0
0	0	0	1	$\alpha_{4,5}$	$\alpha_{4,6}$	$\alpha_{4,7}$	0
0	0	0	0	1	$\alpha_{5,6}$	$\alpha_{5,7}$	$\alpha_{5,8}$
$\alpha_{6\,1}$	0	0	0	0	1	$\alpha_{6,7}$	$\alpha_{6,8}$
$\alpha_{7,1}$	$\alpha_{7,2}$	0	0	0	0	1	$\alpha_{7,8}$
$\alpha_{8,1}$	$\alpha_{8,2}$	$\alpha_{8,3}$	0	0	0	0	1

Fig. 1. Perpetual coding vectors for $g = 8$ and $w = 3$. The α's denote randomly chosen elements from F_q

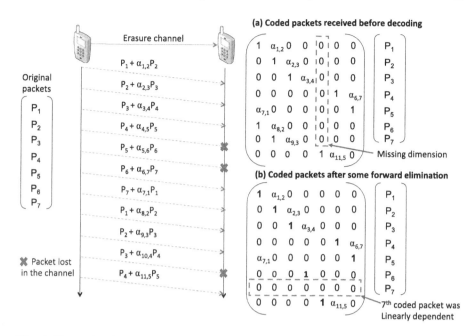

Fig. 2. Example of a transmission process over a packet erasure channel, using perpetual codes.

the receiver gets seven DOF. Due to the erasures on the channel, the 4th and 5th packets are lost. These two coded packets are $P_4 + \alpha_{4,5} \cdot P_5$ and $P_5 + \alpha_{5,6} \cdot P_6$ where P_5 is the common packet between them. P_5 will not be part of any coded packets until the 11th coded packet is received.

Therefore, as we can see in Fig. 2(a), the receiver will be missing a dimension until it receives the 11th coded packet (i.e., packet $P_4 + \alpha_{11,5} \cdot P_5$). With the reception of the 8th coded packet $(P_1 + \alpha_{8,2} \cdot P_2)$, P_1 and P_2 get decoded. Having P_2 decoded leads to decoding P_3, and, consequently, P_4. Moreover, having decoded packet P_1, packets P_7 and P_6 get decoded. Therefore, at the reception of the 9th coded packet, all packets except P_5 are decoded. Moreover, the 9th coded packet $(P_2 + \alpha_{9,5} \cdot P_3)$ is linearly dependent as shown in Fig. 2(b). The transmission of the 10th coded packet fails. Finally, when the 11th coded packet is received, a new dimension is added to the decoding matrix and P_5 can be decoded.

In this example, the source transmits 11 coded packets: three packets are lost, and eight are received by the destination, which successfully decodes the packet generation. As we described before, the receiver gets one linearly dependent packet and seven linearly independent packets. Thus the corresponding *overhead* is $\frac{1}{7}$.

3   Analytical Model

In this section, we first provide a model characterizing the dimension of the subspace defined by the received coding vectors for perpetual codes. Using this model, we analyze the impact of ω on the *overhead* of perpetual codes. Finally, we provide an analysis for number of transmitted packets for perpetual codes in presence of an erasure channel.

Let us consider the use of perpetual codes in the transmission of a generation of g packets. Since the generation size is g, the size of ω is less than g:

$$0 \leq \omega \leq g - 1. \tag{2}$$

We define a class C_i as the set of received coded packets with pivot equal to i. Since we have g different pivots, a coded packet belongs to one of g classes. A class C_i has ω randomly chosen coefficients starting after i^{th} pivot position. We can model the perpetual codes considering that a receiver needs to accumulate linearly independent (L.I.) coded packets by using the different g classes. We define $\dim(C_i)$ as the dimension of the subspace corresponding to class C_i. When the destination receives $\omega + 1$ L.I. coded packets for a class, it can decode that class. Thus, the dimension of the subspace defined by the coding vector of a class can not be greater than $\omega + 1$:

$$0 \leq \dim(C_i) \leq \omega + 1, \forall i \in \{0, \cdots, g - 1\}. \tag{3}$$

To be able to decode a generation, the receiver needs to accumulate g L.I. packets over all classes.

$$0 \leq \dim\left(\bigcup_{i=0}^{g-1} C_i\right) \leq g. \tag{4}$$

For two consecutive classes C_i and $C_{i+1 \mod g}$, the receiver needs to receive $\omega + 2$ coded packets in total to be able to decode both classes.

$$0 \leq \dim(C_{i \mod g} \cup C_{i+1 \mod g}) \leq \min(\omega + 2, g). \tag{5}$$

We can generalize the previous result for classes $C_i, C_{i+1 \mod g}, \cdots, C_{i+k \mod g}$:

$$0 \leq \dim\left(\bigcup_{j=i \mod g}^{i+k \mod g} C_j\right) \leq \min(\omega + k + 1, g) \tag{6}$$

$$\forall i \in \{0, \cdots, g - 1\}, \forall k \in \{1, \cdots, g - 2\}.$$

Consider the subset of k coded packets generated by the last k coding vectors of the coding matrix, represented as a box in Fig. 3. We now show that when the sender transmits k coded packets belonging to k adjacent classes of the box, and $\omega + 1$ coded packets are lost, the sender must transmit, again, at least one packet from one of the k classes in the box, in order to be able to decode all of the packets.

$$
\left\{
\begin{array}{c}
\\ \\ \\ \\ \\ k \\ \\ \\ \\
\end{array}
\right.
\begin{pmatrix}
1 & \alpha_{1,2} & \alpha_{1,3} & \cdots & \cdots & \cdots & \cdots & \cdots & 0 \\
0 & 1 & \alpha_{2,3} & \cdots & \cdots & \cdots & \cdots & \cdots & 0 \\
0 & 0 & 1 & \cdots & \cdots & \cdots & \cdots & \cdots & 0 \\
\cdots & \cdots & \cdots & \cdots & \cdots & \cdots & \cdots & \cdots & 0 \\
\cdots & \cdots & \cdots & \cdots & 1 & \alpha_{i,j} & \cdots & \cdots & 0 \\
\cdots & \cdots & \cdots & \cdots & 0 & 1 & \alpha_{i+1,j+1} & \cdots & 0 \\
\cdots & \cdots & \cdots & \cdots & \cdots & \cdots & \cdots & \cdots & 0 \\
\cdots & \cdots & \cdots & \cdots & 0 & 0 & \cdots & \cdots & 1
\end{pmatrix}
$$

Fig. 3. The k bottom rows represents the k coding vectors where $w+1$ packets created by the box are lost.

In the first transmission round, the sender creates the coded packets from pivot equal to zero to pivot equal to $g-1$ and transmits them to the destination using the first coding matrix. Due to erasures on the channel let's assume $w+1$ coded packets are lost within box, starting from $(g-k)^{\text{th}}$ pivot and ending in $(g-1)^{\text{th}}$ pivot where $k \geq w+1$. Moreover, the first lost packet corresponds to pivot with index $g-k$. These k packets are shown in the box in Fig. 3. We assume that packets corresponding to coding vectors from index 1 to index $g-k-1$ are received successfully. Thus:

$$
\dim\left(\bigcup_{i=0}^{g-k-1} C_i\right) = g - k. \tag{7}
$$

On the other hand, out of k packets in the box, $(k-w-1)$ are received successfully. Therefore:

$$
\dim\left(\bigcup_{i=g-k}^{g-1} C_i\right) = k - w - 1. \tag{8}
$$

Due to the loss of packets in the first transmission round, a second round takes place with coded packets built according to a new coefficient matrix. Since we assume that $g-k-1 > w$, then the first $g-k-1$ coded packets of the second round, assuming no losses occur, will add w DOF to Eq. (7):

$$
\dim\left(\bigcup_{i=0}^{g-k-1} C_i\right) = w + g - k. \tag{9}
$$

Therefore, according to Eq. (6), $C_0, ..., C_{g-k-1}$ get decoded. Moreover, combining Eq. (8) with Eq. (9), the total number of received DOFs becomes:

$$\dim \left(\bigcup_{i=0}^{g-1} C_i \right) = g - 1. \tag{10}$$

According to Eq. (10), the receiver needs one more DOF to be able to decode the full generation. Since all the packets outside the box got decoded, the sender must transmit at least one packet from the box.

The position in the coefficient matrix, of the coding vectors of $w + 1$ lost packets has a huge impact on the number of transmissions. For the case where $k = w + 1$, all coded packets created by the coding vector within the box are lost. These packets correspond to the classes $C_{g-w-1}, \cdots, C_{g-1}$. Therefore, in the second transmission round, the source must transmit at least a packet associated to a coding vector within the box (apart from the first $g - k - 1$ transmissions). Thus, transmissions of the second round encompass, at least, packets from class $C_0, \cdots, C_{g-w-1}$, totalizing $g - w$ coded packets. For the case where $k = w + 2$, the box corresponds to classes $C_{g-w-2}, \cdots, C_{g-1}$. To transmit at least a packet corresponding to a coding vector within the box, the source must transmit coded packets from classes $C_0, \cdots, C_{g-w-2}$, totalizing $g - w - 1$ coded packets. Generalizing, the box comprehends classes $C_{g-k}, \cdots, C_{g-1}$, and the sender must transmit a total of $g - k + 1$ coded packets before transmitting a packet associated to a coding vector from the box.

In the following we derive an approximation to the total number of transmitted packets based on mean field analysis, with the following assumptions:

1. Sender will not transmit more than two rounds. This is a fair assumption for practical system where the loss probability is lower than 0.5;
2. The generation size g is high and the erasure probability ϵ is small;
3. The parameter w is large enough such that, with two transmission rounds, the probability of decoding is high.

Let us define the event IN as: the packet corresponding to the first coding vector of the box of size k, is lost; and w more packets inside the box are lost (and, consequently, $k - (w+1)$ packets are received successfully). The probability of this event is:

$$P(IN) = \binom{k-1}{w} \epsilon^{w+1} (1 - \epsilon)^{k-(w+1)}. \tag{11}$$

When i packets outside the box are lost, the sender must transmit at least $i + w$ packets, in order to recover from these losses. Let's define the event OUT as: losing i packets outside the box. The probability of this event is:

$$P(OUT) = \binom{g-k}{i} \epsilon^i (1 - \epsilon)^{(g-k-i)}. \tag{12}$$

As we mentioned before, the sender must transmit at least $g - k + 1$ coded packets before transmitting a packet associated to a coding vector inside the box. The total number of transmissions when i coded packets outside the box, and $w + 1$ coded packets inside the box are lost is $\max(g - k + 1, w + i)$. To recover from these losses, the sender should transmit $\frac{\max(g-k+1,\omega+i)}{(1-\epsilon)}$ to compensate the packet erasure probability on the channel. Since we are only considering two transmission rounds, the sender can transmit, at most, g coded packets (the maximum possible in the second round). Therefore, the number of transmitted packets is $\min\left(\frac{\max(g-k+1,\omega+i)}{(1-\epsilon)}, g\right)$.

Considering Eqs. (11) and (12), the expected number of transmissions of the second round can be approximated by:

$$T_2 \approx \sum_{k=\omega+1}^{g} \sum_{i=0}^{g-k} \min\left(\frac{\max(g-k+1,\omega+i)}{(1-\epsilon)}, g\right)\binom{k-1}{\omega}\epsilon^{\omega+1}(1-\epsilon)^{k-(\omega+1)}\binom{g-k}{i}\epsilon^{i}(1-\epsilon)^{(g-k-i)}.$$

$$(13)$$

Since the number of packets transmitted in the first round is $T_1 = g$, the total number of transmissions for both rounds is:

$$T_{1,2} \approx T_2 + g. \qquad (14)$$

Adding to $T_{1,2}$ the transmissions of the case when the number of lost packets is less than ω we get:

$$T_T \approx T_{1,2} + \sum_{i=0}^{w} \min\left(\frac{i}{(1-\epsilon)}, g\right)\binom{g}{i}\epsilon^{i}(1-\epsilon)^{(g-i)}. \qquad (15)$$

Equation (15) is an approximation to the total number of transmitted coded packets T_T in two transmission rounds, as function of the generation size g, the *with* ω in perpetual codes, and the channel erasure probability ϵ, under the assumptions stated above. The parameter ω has a big impact in the number of required transmissions using perpetual codes. With small ω, losses on the first transmission round lead to linear dependent transmissions of the second round.

4   Numerical Results

In this section, we first compare the overhead of perpetual codes with the one of RLNC. Then, we validate our analysis with numerical simulations, showing that they are in agreement. Finally, we show that ω is the critical parameter for the overhead of perpetual codes.

To illustrate the overhand of perpetual codes, we performed a simulation study, comparing RLNC with perpetual codes, using the KODO library [9]. Figure 4(a) shows the total number of transmissions for different erasure probabilities for RLNC and perpetual codes, for $g = 256$ and $\omega = 5$. In this example, we assume that the sender can perform multiple transmission rounds, stopping when the receiver fully decodes the generation. However, the other plots only consider two transmission rounds, which is the case we consider in our analysis. As it is shown in Fig. 4(a), even with small packet erasure probabilities, the number of transmitted packets using perpetual codes is close to 500, being almost

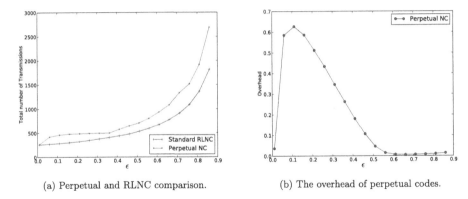

(a) Perpetual and RLNC comparison.

(b) The overhead of perpetual codes.

Fig. 4. The total number of transmissions and overhead for perpetual codes with different error probabilities on the link when $g = 256$ and $\omega = 5$.

50% more than the number of transmissions using RLNC. Figure 4(b) illustrates the overhead of perpetual codes for two rounds when $g = 256$ and $\omega = 5$. For small erasure probabilities, the *overhead* of perpetual codes is almost 0.7. This means that for every DOF received, the system needs to transmit a total of 1.7 packets in average.

Figure 5(a) compares the simulation of two transmission rounds, using perpetual codes, with the analytical results derived in the previous section. The simulation results show that the expression for the number of transmissions of perpetual codes (derived in the previous section) constitutes a good approximation.

Figure 5(b) presents the analytical results on the number of transmitted packets using perpetual codes, for a packet erasure probability of 0.2, $g = 256$, and for different ω values. The figure shows that, for values of $\omega \leq 40$, the parameter ω has a significant impact on the number of transmissions. Moreover, we can see that for $\omega \geq 60$, the impact of ω on the number of transmissions becomes marginal.

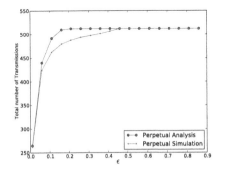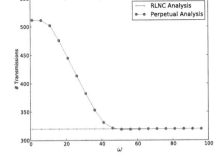

(a) Analysis and simulation results comparison when $\omega = 5$.

(b) Analysis results for different ω values.

Fig. 5. The total number of transmissions for perpetual codes with different error probabilities on the link when $g = 256$.

5   Conclusion

In this paper, we proposed a mathematical model describing behavior of perpetual codes for transmissions over a packet erasure channel. Using this model, we studied the impact of ω on the transmission *overhead* of perpetual codes. We show that, over packet erasure channels, using perpetual codes with small ω leads to the reception of linearly dependent packets. The corresponding overhead can reach values close to 70%, for low erasure probabilities and for $\omega < 10$. In order to mitigate the overhead, the sender must do an appropriate choice for the value of ω. A proper choice for ω is highly dependent on the channel erasure probalility. We derived an expression that is an approximation of relationship between ω and the number of transmitted packets. Moreover, we performed numerical simulations of the perpetual codes using the KODO library. The simulation results highlight the effect of different ω values on the number of transmitted packets and on the *overhead*. They show that our analysis provide a good approximation for the number of transmitted packets using perpetual codes. However, we have only considered the systematic pivot selection, meaning that the pivot elements are chosen sequentially, one after the other. As future work, we will provide an analysis for the case where the pivot elements are drawn at random. Furthermore, we will consider the impact on the overhead of error bursts in the channel.

References

1. Ahlswede, R., Cai, N., Li, S.Y., Yeung, R.W.: Network information flow. IEEE Trans. Inf. Theor. **46**(4), 1204–1216 (2000)
2. Heide, J., Pedersen, M.V., Fitzek, F.H.P., Medard, M.: A perpetual code for network coding. In: 2014 IEEE 79th Vehicular Technology Conference (VTC Spring), pp. 1–6, May 2014
3. Heide, J., Pedersen, M., Fitzek, F., Larsen, T.: Network Coding in the Real World, 1st edn, pp. 87–114. Academic Press, Cambridge (2011)
4. Ho, T., Medard, M., Koetter, R., Karger, D.R., Effros, M., Shi, J., Leong, B.: A random linear network coding approach to multicast. IEEE Trans. Inf. Theor. **52**(10), 4413–4430 (2006)
5. Li, S.Y.R., Yeung, R.W., Cai, N.: Linear network coding. IEEE Trans. Inf. Theor. **49**(2), 371–381 (2003)
6. Luby, M.: Lt codes. In: Proceedings of the 43rd Annual IEEE Symposium on Foundations of Computer Science, pp. 271–280 (2002)
7. Maymounkov., P.: Perpetual codes: cachefriendly coding. Unpublished draft. http://pdos.csail.mit.edu/?petar/papers/maymounkov-perpetual.ps
8. Nistor, M., Lucani, D.E., Barros, J.: Hardware abstraction and protocol optimization for coded sensor networks. IEEE/ACM Trans. Netw. **23**(3), 866–879 (2015)
9. Pedersen, M.V., Heide, J., Fitzek, F.H.: KODO: an open and research oriented network coding library. In: Workshop on Network Coding Applications and Protocols (NC-Pro), Spain (2011)
10. Shokrollahi, A.: Raptor codes. IEEE Trans. Inf. Theor. **52**(6), 2551–2567 (2006)

Author Index

Printed in the United States
By Bookmasters